CUTTING THE CORD

The Cell Phone Has Transformed Humanity

Martin Cooper

RosettaBooks®
NEW YORK 2020

Cutting the Cord: The Cell Phone Has Transformed Humanity

Copyright © 2020 by Martin Cooper

First edition published 2020 by RosettaBooks

Jacket design by Rupa Limbu and Lon Kirschner

Additional designs by Alexia Garaventa

Cover photograph by Serge Hoeltschi

Author photo by David Friedman

ISBN-13 (print): 978-1-9481-2274-0
ISBN-13 (ebook): 978-0-7953-5302-4

Library of Congress Cataloging-in-Publication Data:
Names: Cooper, Martin, 1928- author.
Title: Cutting the cord : the cell phone has transformed humanity / Martin Cooper.
Description: New York : RosettaBooks, 2020. | Includes bibliographical references and index.
Identifiers: LCCN 2020037807 (print) | LCCN 2020037808 (ebook) |
ISBN 9781948122740 (hardcover) | ISBN 9780795353024 (ebook)
Subjects: LCSH: Cell phone systems—History. | Cell phone systems—
Social aspects. | Telematics—History.
Classification: LCC TK5103.2 .C657 2020 (print) | LCC TK5103.2 (ebook) |
DDC 384.5/34—dc23
LC record available at https://lccn.loc.gov/2020037807

RosettaBooks®
www.RosettaBooks.com
Printed in Canada

"Each friend represents a world in us, a world possibly not born until they arrive; and it is only by this meeting that a new world is born."

—Anaïs Nin

"At first they could not understand it at all; but presently Shaggy suspected the truth, and believing that Ozma was now taking an interest in the party he drew from his pocket a tiny instrument which he placed against his ear.

"Ozma, observing this action in her Magic Picture, at once caught up a similar instrument from a table beside her and held it to her own ear. The two instruments recorded the same delicate vibrations of sound and formed a wireless telephone, an invention of the Wizard. Those separated by any distance were thus enabled to converse together with perfect ease and without any wire connection."

—The Tik-Tok of Oz, by Frank Baum (1914)

To the First Lady of wireless
and
the only lady in my heart

Arlene Harris

CONTENTS

PREFACE

J ust as I was completing the manuscript for this book, the COVID-19 pandemic struck the world. By the time the book is published, it's highly probable that the first wave of the pandemic will have passed. Individuals and societies will be attempting to return to normal—or find a new normal.

Already, in the early stages of the pandemic and its accompanying economic and social lockdowns, there have been noticeable changes regarding cell phone use. On the day I write these words, a headline in the *New York Times* reads, "The Humble Phone Call Has Made a Comeback."[1]

Verizon customers were making "an average of 800 million wireless calls a day during the week, more than double the number made on Mother's Day." Likewise, AT&T reported that "the number of cellular calls had risen 35 percent" during the pandemic. The calls, moreover, were 33 percent longer than the pre-crisis average.

Stuck at home, confined to apartments, and unable to interact in person with others, people were picking up the phone. "We are craving human voice," my friend Jessica Rosenworcel, FCC commissioner, told the *Times*.

I'm not surprised. There is something personal about hearing a person's voice that is lost over email and text message. (Text messages also rose by a third during the same April week, to more than nine billion sent per day by Verizon customers. No doubt some number of those were attempts to coordinate all the extra phone calls.)

What really struck me about this news report was that the increase in phone calls pertained to cell phones. Landlines—or wired phones—have been in long-term decline. There are 90 million fewer today than there were in 2000.[2]

This is one way to mark the completeness of the cell phone takeover: we're making more phone calls to each other, but not with the technology that defined communications for a century. That, however, is a backward-facing milestone. The coronavirus pandemic will usher in other permanent changes to our relationship with cell phones.

Millions of people, for example, are learning how easy it is to make a video call on their cell phone. Grandparents are using the Houseparty app to hang out and play games with the grandchildren they can't hug. Friends are using FaceTime and Skype to host virtual happy hours with each other. Employers who previously shunned remote work tools are finding that Zoom and Teams will become essential to future operations.

Privacy concerns are bowing to the need to engage in contact tracing in order to contain the virus. East Asian countries have been ahead of the West in using cell phones to monitor movement and interaction. Just before I wrote this, though, Apple and Google announced their collaboration to develop contact-tracing software. By the time you read this, your cell phone may already be able to tell you if you've been exposed to someone with COVID-19. You can be sure that this type of development—even if announced to be temporary or confined just to coronavirus—will presage more permanent changes. A toe will have been inserted into the privacy door that is unlikely to ever be removed.

The pandemic will also accelerate the use of cell phones in health care. Telemedicine visits boomed during the crisis, with many of those "calls" taken on phones. Cell phones will also become a more widespread and accepted way to monitor vital signs and potential disease symptoms.

Schools of all types as well as other organizations are gaining extended experience with operating at a distance. Online and distance learning have been steadily gaining steam for several years—the pandemic turbocharges this. It remains to be seen, at the time of this writing, whether parents and students will feel comfortable with on-campus living and learning. The concentrated experience of the pandemic will accelerate improvements in remote classes, virtual conferences, and more. Everyone is learning how to make them work better, and quality is undoubtedly already increasing.

We will also see permanent shifts in exchanges of different kinds. Physical business cards will disappear, replaced by virtual exchange through Near Field Communication between phones. A cashless society will also become reality: physical currency has long been known to be a transmission source for bacteria and viruses. How many people will voluntarily choose to pay with a twenty-dollar bill—and how many merchants will prefer cash—over the credit card stored in their phone that they can simply wave in front of a scanner?

More fundamentally, the coronavirus pandemic and resulting social changes highlight two observations that I will make repeatedly throughout this book. First, people are mobile. Second, people connect with people, not places. To me these are as basic to human existence as our physical needs. The lockdowns and social distancing brought on by the virus reinforced their truth. In the first instance, by restricting the very mobility we all take for granted. And, in the second, by reminding us all that we need each other and, if we can't sit next to one another, a phone call is the next best thing.

INTRODUCTION

The crash of lightning broke my reverie.

It's April 3, 2013, forty years to the day since I made the first public cell phone call on a New York City street. I am sitting in a rickety old chair at an equally worn desk staring out the window at an unremarkable hill. In my imagination, I chat with Guglielmo Marconi 120 years ago, as I sit in his chair, in his laboratory, preserved in a small town just southwest of Bologna, Italy. I ask, "Signor Marconi, whatever gave you the idea that you could send a radio signal from this room to the other side of Celestini hill?"

You see, the chair, the window, and the hill are more remarkable than they seem at first glance.

"You're not the first person to ask," I imagine Marconi responding. "I had, for years, been dreaming of sending electromagnetic waves over long distances. I know that the earth is round and there are mountains and man-made objects that can separate a receiver and a transmitter. I needed to know whether electrical waves could travel through or around such objects."

Marconi was driven by very basic curiosity.

"The experiment was very simple," he continues. "My assistants stood on the other side of Celestini hill with a coherer and a rifle.[1] If the coherer detected the signal that I was sending from my workroom, one of the assistants, my brother, was to fire the rifle. Which he did!"

My imagined conversation with Marconi is interrupted by the lightning over Celestini hill. I blink and look around. Three dozen local officials, technology executives, and friends crowd around me as I sit in Marconi's lab in the Villa Griffone, his childhood home, now the Museo Marconi. My daughter, Lisa, is also with me, as is Princess Elettra Marconi, daughter of the great radio pioneer. Her father was one of the first people to send a wireless signal across long distances, including the Atlantic Ocean. He made radio communications practical.

Elettra later told me she was named after the lavish yacht Marconi bought himself after his inventions began to pay off financially. The boat was outfitted with stem-to-stern radio, direction finder, and radar antennae. Her father was very proud of the newly invented radar. Apparently, he once had her mother cover the ship's pilot house windows with sheets so that he could demonstrate the ability to navigate through a dangerously reefed area using only radar.

And now here I am, being asked to sit in Marconi's original chair, recipient of a prize in his name. The Marconi Prize is awarded each year to innovators who helped advance information and communications technology. Recipients have included Tim Berners-Lee, creator of the World Wide Web, and the cofounders of Google. Vint Cerf, a "father" of the internet (for the TCP/IP protocol developed in 1973, the same year as our first cell phone at Motorola) is another recipient.[2] As a quick aside, it was he who sent me the charming passage from Frank Baum's *The Tik-Tok of Oz* that appears at the beginning of this book. Vint included a note that read, "The Wizard would be Marty Cooper!" That is too kind. While I envisioned the first handheld cell phone, prompting some to see me as the Wizard, it's also true that no wizard works alone. I was supported by the work of many others, including Bob Galvin, longtime CEO of Motorola, who received the Marconi Society Lifetime Achievement Award in 2011 (the year he died).

How did I, someone who always assumed he was the least knowledgeable and experienced in any room, end up in the same company as these remarkable names? I would rather sit in this chair and gaze out the window, imagining myself in deep conversation with Marconi. I know it sounds presumptuous, but I like to think we had a lot in common. We both depended a lot on luck, dreaming, and persistence.

An Italian journalist asks, "Dr. Cooper, can you tell us about the innumerable challenges you had to overcome to invent the cell phone—and how you overcame them?"

Just thinking about the last months of 1972 and the early part of 1973 sends a tingle up my spine.

"I was one of many," I clarify. "I did dream up the world's first handheld, portable cell phone, but it took a team of skilled and energetic people to build that phone and make it work. And thousands more executives, engineers, and marketers to create today's trillion-dollar industry."[3]

People raise their cell phones to take pictures. I chuckle to myself. In some countries I may be celebrated as "el padre del teléfono móvil," but I certainly didn't

*Me, sitting at Guglielmo Marconi's desk where he conducted some
of the first wireless experiments in the 1890s.*

imagine there would someday be a phone with a built-in camera. Never mind a
powerful computer, internet access, Wi-Fi, Bluetooth, and a slew of sensors.

It's difficult for many of us to remember—and many simply don't know—that for
most of the twentieth century the only way to place telephone calls was through
the Bell System. The Bell System was AT&T and its associated organizations,
Bell Labs and Western Electric, which dominated telecommunications through
a government-regulated monopoly. It was known as "Ma Bell," a name that captured how dominatingly pervasive the company was.

For twenty years, through the 1960s and 1970s, I was at the center of a battle
with AT&T over the future of how Americans would communicate. At the heart
of that battle was the introduction of the first handheld, portable cell phone—
one on which I made the first public call.

The invention and commercialization of the cell phone set off a frenzy of
entrepreneurship in the United States and other countries.[4] I was part of that

With Princess Elettra Marconi, daughter of the great radio pioneer, during my visit to accept the Marconi Prize.

frenzy, too, and have spent most of my career working on portable, wireless communications. After thirty years as a corporate employee at Motorola, I shifted gears and became an entrepreneur. With various partners—including my wife, Arlene Harris—I helped start Cellular Business Systems, Cellular Pay Phone, SOS Wireless, GreatCall, and ArrayComm.

This book tells the story of how the cell phone was developed. But this is not a historical document, nor is it an autobiography. Rather, this book is partly a memoir based on my recollection of events that changed how people communicate. It's the story of how a dreamer helped launch the cell phone, what led to its creation, and how it has transformed society. Along the way, I learned a good deal about business, innovation, strategy, and more. I'll share some of those stories, too.

Most importantly, this book is about the future. Although the technology embedded in the modern cell phone is phenomenal, our ability to adapt to that technology is still in early days. Humanity has realized only a small fraction of the potential of the cell phone to help solve the big problems of society, including poverty and disease. The productivity improvements stimulated by the connectivity of the cell phone can help reduce the gap between the haves and the have-nots that is, I believe, the source of most conflicts in the world.

The first cell phone was designed and assembled in a bit over ninety days. Yet the principles and insights informing that process had accumulated over two decades of prior work on portable and wireless communications. Over the subsequent forty years of my career, I would find myself extending and applying those principles repeatedly—and I'm still learning.

Like anybody, I have made a huge number of mistakes during my career. I've tried to learn from those mistakes to avoid repeating them. But I have also tried to forget the unpleasantness of those mistakes and relive the few successes I've enjoyed—I'm only human, after all.

Above all, I hope this story inspires others in the same way that Marconi's story inspired me. He isn't generally considered to be the inventor of the best radio technology—that would be Nikola Tesla (at least in the United States). Marconi also wasn't necessarily a scientist in the traditional sense. He was looked down upon by the more theoretically minded and even called himself an "ardent amateur student of electricity."[5] Marconi loved to experiment with sending electric signals through the air—but he was driven by a bold vision.

In the 1890s, Marconi stared out the window over his desk at Celestini hill and asked if it would block the wireless Morse code messages he tapped out on the brass and wooden receiver that still sits on his desk in the Museo.[6] When his brother fired the rifle, he heard the signal of success.[7]

This emboldened Marconi to make further experiments, sending wireless signals over increasingly long distances. By 1901, he had sent a wireless message across the Atlantic Ocean. In 1909, he was awarded the Nobel Prize in Physics. Yet this highest honor of scientific achievement was not the culmination of Marconi's bold vision. He was the ultimate entrepreneur, seeing the great potential of wireless signals to reshape how people connected and communicated with each other. His interest was in getting wireless communications into practical use, in commercializing radio. Yes, he thrived on the prizes and public adulation, but his passion, his obsession, was making radio practical and profitable. His vision and the organization that resulted—the Wireless Telegraph & Signal Company, which grew into a global communications company—became the basis of a new and important industry.

His vision was put to the test when the *Titanic* sank in 1912. Nearly everyone on the ship might have died without Marconi's two on-board telegraph operators transmitting wireless SOS signals over long distances. Their signals summoned rescue crews who were able to help save more than seven hundred lives.[8]

As I stare at Celestini hill, I can draw a line through the last century from Marconi's early experiments and commercial successes to the array of mobile phones taking my picture here in Marconi's lab. The line is not straight, and it nearly veered off course at many moments, especially when threatened by the monopolists at AT&T.

The Marconi Society, which awards the Marconi Prize, says it recognizes "exemplary" and "significant scientific" contributions that change people's lives and create new industries. This moment is tremendously humbling and, for an electrical engineer who spent years dreaming of ways to expand *portable* wireless communications, positively dreamlike. Which is fitting: I've spent most of my life dreaming of how to do *everything* differently. When I was twelve, I dreamed of a transcontinental tunnel using magnetic levitation to power trains through a vacuum. I wanted to use it to safely shuttle people between Chicago and Los Angeles. I dreamed of ways to enable undersea living and deep-space rocket ships.

These fired my imagination, but they were dreams divorced from any practical way to make them real. Dreaming fueled my curiosity to understand how things work—but I was determined to gain that understanding firsthand, not merely in my head. So, as a boy, I built model airplanes. When I was a bit older, I enjoyed disassembling my Triumph Stag convertible when others would have discarded it as the most unreliable car ever created. Still later, when I was a naval officer in training, I snuck onto the deck during a driving rainstorm to help sailors haul mooring line. I refused to stay below and imagine what was going on. I had to experience it.

I also dreamed about a new communications device, one unburdened by the telephone wires that, for over a hundred years, had imprisoned people in their homes and offices when they wanted to connect with others at a distance. That dream is what gave me the honor to sit in Marconi's chair.

The handheld, portable cell phone is no longer recognized as very revolutionary. After all, everybody has one. Yet that revolution has barely started. The portable telephone industry is still in its infancy; we're just starting to learn the important benefits of connecting people and machines.

Portability is the *people* imperative. The ability of *individuals* to communicate directly with each other, at any time and wherever they are, changed the world.

The creation of the first truly portable phone—in pursuit of enabling communication between people independent of place—was not a solitary pursuit. Developing the cell phone was a group effort, and it built upon a strong legacy of

Reunion of the Motorola team that developed the DynaTAC—the world's first handheld,
portable cellular phone. October 2007.

innovation and inspiration at Motorola, the company where I worked for nearly three decades. Invention doesn't happen by sitting alone under an apple tree and waiting for an idea to drop. Breakthroughs happen in environments that are conducive to it, where dissatisfaction with the status quo is the norm, where people work together to solve common challenges. Where objectivity is prized and politics eschewed. Those elements defined Motorola.

"Reach out! Do not fear failure!" That was the exhortation of Paul Galvin, Motorola's founder. Paul and his son, Bob, created an environment that drove us to ceaselessly seek ways to improve the way people communicated. We were never satisfied, and failures were understood to be part of the process. My colleagues and I saw a better world enhanced by radio waves and did our best to execute that vision. I will always remember and be inspired by Bob Galvin, Bill Weisz, and John Mitchell, each of whom was an exceptional role model. I can't imagine having achieved anything without that inspiration and their tolerance for my weaknesses, like my chronic disorganization and my inclination to speed ahead without adequate thought or preparation.

My dream of a new communications device found fertile ground in Motorola's environment, and I pushed it tirelessly. My passion was matched by the improbability of my dream. We were confronting not only Ma Bell, one of the most powerful organizations of the twentieth century, but also rampant skepticism.

Hindsight can make many ideas seem obvious, and that's certainly how the cell phone looks today, when there are billions in everyday use. In the 1970s and 1980s, however, all right-thinking people "knew" that there was no market for handheld, portable phones.

What I concluded from all this skepticism was that not everyone knows how to dream. I dream so often about the future I frequently think I live there. For me, the future is always so much more interesting and exciting than the past. When it comes to the cell phone, this is doubly true. In the nearly forty years since the dawn of the first commercial cellular systems, the mobile phone has had an incalculable impact on humanity.

But the real fun comes from even more dreaming—dreaming about the continued transformative impact the cell phone will have on our lives. It is already affecting how we care for ourselves, how we work together, and how we learn. It has contributed to dramatic reductions in poverty in many countries. We are only at the beginning. The cell phone will reshape health care; it will make education unrecognizable. It will help eliminate poverty.

But not without more dreamers and more organizations that allow those dreamers to try, fail, and try again. And not without public policy that promotes competition and effectively manages use of the radio frequency spectrum.

I hope that this book, and my story, will inspire you to take up these challenges: What will you dream, how will you extend the benefits of the cell phone—and wireless, portable communications in general—for more and more people?

PART I

OUT OF THE DITCH

Immigrants and Entrepreneurs

In 1919, amid the Ukrainian War of Independence and the Russian Civil War, Cossacks on horseback galloped through the village of Pavoloch in Ukraine killing townspeople at random. A fourteen-year-old girl dived into a ditch, narrowly avoiding the slice of a Cossack's saber.

This was merely an advance party: more Cossacks were on the way, and they would plunder the town, killing, raping, and wounding as they went. Nearly 90 percent of Pavoloch was Jewish, and it would be the site of one of the periodic but horrific pogroms aimed at Jews in Ukraine in those years.

The girl ran home to warn her family. Her father, Nathan Bassovsky, made an immediate decision. Pogroms had devastated other Jewish towns the previous year, nearly wiping them out—he would not let his wife and six children fall victim to another one. Before more Cossacks could arrive, Bassovsky organized and financed a wagon train and invited his neighbors and friends to join in an escape. Those who accepted loaded everything they could onto horse-drawn wagons and left.

It took months for the caravan to cross Europe. Partway, they sold the horses and wagons and traveled by train to Antwerp, Belgium. They were not alone. The Belgians had created a facility where thousands of pogrom victims could bathe, launder their clothes, and leave Europe.

The Bassovsky family.

Nathan, with his eldest son, Morris, departed ahead of the rest of the family for Canada, settling in Winnipeg among others who were welcomed there by organizations responding to the horror in Europe. In 1921, the fourteen-year-old girl, Mindel Bassovsky, and her two sisters, Birdie and Rose, and the remaining brothers, Max and Frank, boarded the SS *Caronia* bound for Halifax. The five children were among millions of people who fled Europe for Canada and other countries.

The Canadian National Railroad document shows that Mindel arrived in Canada on August 13, 1921. She declared that she had the minimum amount of fifty dollars required for entry and that she was on her way to join her father in Winnipeg. On the line that asked, "Nearest relative in country from which you came," she wrote, "nobody." All Mindel's immediate family had abandoned Ukraine, and she knew what fate faced the Jews who remained there.

Mindel Bassovsky was my mother; her perseverance and sharpness were impressed upon me at an early age. By the mid-1920s, Mindel had met and married my father, Osher Kuperman, later Arthur Cooper, who had also arrived in Canada in 1921 on the SS *Antonia*, traveling from Skvyra, Ukraine. On his Canadian

Entry document for Mindel Bassovsky into Canada, 1921.

The SS Antonia. *The ship my father traveled on from Antwerp to Halifax in 1921.*

entry card, he wrote that he was "joining cousins" in Canada. My mother also changed her name, to Mary.

Sometime in the late 1920s, my parents emigrated again, this time to Chicago. I was born there in 1928. Not long after that, they returned to Winnipeg, and my brother was born there in 1932.

Their first attempt to become financially independent was a laundry business in Chicago, which they had the opportunity to purchase. The business was expensive, so my parents wanted to be sure it was a good investment. The owners let them work at the laundry for a one-week trial period. Business was booming; my parents turned over their savings and bought the laundry.

Sadly, it was a Potemkin village, a sales charade meant to impress, and deceive, my parents. The once steady stream of customers evaporated pretty much the morning my parents opened for business. Evidently, the previous owners had organized their large and extended family to pose as customers during the trial week. That prompted my parents to return to Winnipeg, packing up their Chicago belongings—including their young son.

In Winnipeg, my parents owned and operated a modest grocery store near Main Street at Redwood and Charles. It was in a single-story wood building, and we lived in the rooms behind the store. We had a small yard in the back with a single tree, where I built a "tree house" when I was six or seven—just a few wooden boards nailed into the branches.

The store itself was a single room lined with shelves, with more freestanding shelves in the center. The most memorable thing for me was the corn flakes. I remember boxes of them stacked on the high shelves. Too high and too close, though, to the stovepipe along the ceiling that took out fire exhaust from the small, coal-fired stove that heated the store. No one was hurt when the inevitable fire broke out—unless you count the extended period during which we all ate lightly blackened corn flakes for breakfast. The store, luckily, did not burn down.

Unsurprisingly, many of my childhood memories of Winnipeg involve ice and snow. We skated on the streets when they became sheets of ice during the winter. Our neighbors across the street flooded their backyard to create an ice rink that all the neighborhood kids used well into the spring. Milkmen would convert their horse-drawn delivery wagons into sleighs. Pavement was a rare sight until the big spring melt.

Around 1937, after the grocery store fire, our family moved to Fort William, Ontario (which later became part of Thunder Bay). There, my parents started another grocery store. I was now a voracious reader and a nerd, the last to be chosen when teams were picked for schoolyard games. After a year or so in Fort William, with the grocery store failing, we found ourselves back in Winnipeg, but just for a short time. My parents were planning a move back to Chicago.

In 1937, my father hopped on a freight train to cross the US border illegally. I desperately wanted to join him, but that wasn't an option. My mother and my

My maternal grandparents.

Me (with curls!) in Winnipeg in the 1930s.

brother, Will, and I went to Niagara, Ontario. We stayed overnight in the cheapest hotel in town and, the next morning, walked across a bridge into the United States. When the border official asked for the purpose of our visit, my mother replied, "We're going shopping." It took four years for us to return from that shopping trip.

My parents and brother were "illegal" immigrants. I had been born in the United States so was a citizen. My nephew Steve, digging through old family records, found a US border entry card showing that my mother, father, and brother legally re-entered the United States in 1943. A charitable organization in Chicago assisted with the naturalization process, and they became citizens in 1945.

My folks found an inexpensive flat on the west side of Chicago that was our home for about five years. My father got a job in a luggage factory. My mother, following her entrepreneurial instinct, sought out business opportunities where she could use her sales skills productively. I still recall my embarrassment when I accompanied her as she sold corsets to chubby ladies in a rented hotel room on Michigan Avenue.

She eventually discovered the installment sales business. My mother was a dynamo, a woman incapable of walking slowly, and this job fit her perfectly. She would take a bus to a distant neighborhood and, with a throw rug under her arm, knock on one door after another, offering the rug for sale for as little as fifty cents a week. She returned each week to collect payment and always with a new household offering. Wholesalers of furniture, clothing, and household goods gave independent entrepreneurs like my mother credit to sell their products. The entrepreneurs extended credit, in turn, to their customers. If only the internet existed then, she probably would have founded Amazon, and I would have been the snotty rich kid who never had to work a day in his life.

My father later joined her in the business. The luggage factory was the first and last time he ever worked for someone else. Both of my parents were what today we call serial entrepreneurs. In fact, among my mother and her five siblings, only Rose's husband worked for someone else—it was not overtly mentioned, but Harry was the odd man out in our family.

My mother had an irresistible personality and was superb at cold house calls; my father hated them. She would acquire new clients, and he would take on and develop the account. In his quiet way and with a healthy sense of humor, he turned out to be a successful salesman. That was how they supported our family for the rest of their working lives. They made enough to purchase cars so that they could extend their customer base to the suburbs. Their customers became

My father's 1943 border entry card for the United States.

almost like family members; they attended our weddings and celebrations and we attended theirs.

For a teenage boy, my mother's personality could be embarrassing. She was always getting into conversations with strangers. Yet I find myself doing the same. Like her, I talk to strangers and develop friendships anywhere. I realize she must have passed along her love of people to me. Like her, I don't know how to walk slowly. And, like her, I've always been selling, but in my case, I sell ideas and dreams.

Our family wasn't wealthy, but I don't ever remember going hungry during the Depression. My brother and I shared a bed for several years, and we even had a boarder for some time. My parents did well enough in their business ventures to afford an early television set and a new Plymouth car.

Even in my earliest memories, I had an intense interest in how things worked. I *knew* that, someday, I would be an engineer. As a five-year-old in Winnipeg on the cracked and uneven concrete sidewalk in front of our grocery store, I observed some boys using the sun's rays through a magnifying glass to burn a piece of paper. I spent hours trying to replicate the experiment. Despite the use of Coke bottle bottoms and pieces of glass, I was unsuccessful even after I heated the glass with matches that I took from the store.

In the basement of our apartment on Maypole Avenue in Chicago, using money from part-time jobs, I constructed a photographic darkroom and became a pretty good photographer. I lobbied our landlord to let me install a TV antenna on the roof. He refused, having no idea what an antenna was and fearing anything modern, but I installed it anyway and he never noticed.

My fascination with how things worked drew me into a love affair with cars. These were machines I could try to understand by taking them apart and putting them back together. Like many of my childhood and teenage friends, I could tell you every detail of every car model and how they changed through the years. I didn't get a car of my own until 1952. It was a beautiful 1949 Packard convertible, its whitewall tires a striking contrast with its gleaming black body. It wasn't a great car. The automatic clutch never worked right, and it had awful suspension

Me with my beautiful but terribly unreliable 1949 Packard.

that caused the car to shudder terribly at the slightest bump in the road. But my convertible elevated my bachelorhood to heights I never envisioned.

It was the deficiencies in cars that allowed me to learn about them. Most cars in the 1940s and 1950s were fundamentally unreliable. The greatest deficiency was mine, though; I could not resist a beautiful car and as a result had plenty of opportunities to develop skills in car repair. I started with details like adjusting the thermostatic spring on the carburetor that, at least theoretically, acted as an automatic choke. My expertise in car repair grew through the years as I successively acquired cars that were arguably the most unreliable ever built, including a Triumph Stag, a Jaguar XKE and a couple of Jaguar XKSes. The English created the most beautiful and least reliable cars on the road. I was captivated by their beauty and saw their faultiness as a chance to take them apart to get to know them better.

In Chicago, I attended school beginning in the fifth grade and later went to Crane Technical High School. Only in later years did I realize that Crane was a trade school. Every semester, in addition to the traditional liberal arts and science courses, we did hands-on learning in workshops, starting with woodshop and graduating to forging, foundry, printing, machine, and electrical shops. The confidence I built in these shops supported my entrepreneurial bets for the rest of my life.

⌒✳⌒

Despite our modest circumstances, there was never any doubt that I would go to college. Tuition at the Illinois Institute of Technology (IIT) in 1946 was $128 per semester, almost $1,700 in today's dollars. When I graduated from high school in January of 1946, I lived at home and commuted to IIT.

Although I never considered any school other than IIT, it turned out to be perfect for me. I belonged to a fraternity and two honorary groups and had a close-knit circle of friends. The school's focus was on imparting a serious and practical engineering education. Homework assignments were tough, teaching superb, and parties rare. Classes were held in converted tenement buildings and repurposed Quonset huts left over from World War II. I felt privileged to be there and have been grateful ever since.

Still, I felt like a burden on my parents and looked for other ways to fund my education. One of those was an offer from the US Navy. They would pay for my tuition, books, and incidentals in exchange for participation in annual two-month summer cruises over the next three years and a commitment to active military service for three years after graduation. I jumped at the opportunity, happy to relieve my parents from the struggle of paying for my university education.

It was an extraordinary deal; I got much more than financial support out of it. The navy helped me grow up. My time in the military taught me about leadership, responsibility, and getting along with people.

The summer cruises were a blast. I spent them on a variety of navy ships including the aircraft carrier USS *Boxer*—where I had my first experience flying in a propeller plane—and the heavy cruiser USS *Helena*, visiting places like Guantánamo Bay and the Panama Canal. My commitment also included time at the Naval Training Station in Waukegan, Illinois.

One night onboard the *Helena*, cruising somewhere in the Pacific Ocean, we were on the aft deck watching a movie, *The Boy with Green Hair*.[1] An unbearable pain in my stomach developed, so I headed for the sick bay. It seemed like it was miles away and, with the pain intensifying, I had to crawl the last hundred feet. The pharmacist-mate on duty assigned me a bed, where I lay for what felt like hours until the duty doctor showed up. He told me my appendix was about to burst and arranged an immediate operation. While the pharmacist-mate prepped me, he explained to me that the doctor was an alcoholic, but that I shouldn't worry since he took drugs to steady his hands for operations! To minimize the pitching and rolling of the *Helena* during my operation, the entire fleet was turned off course and into the wind, helping calm the swells. To calm myself, since I was given only a local anesthetic, I entertained the surgical crew with my limited repertoire of jokes while watching the operation in the mirror surface of the ceiling. Thankfully, it was a success and I was back on duty in a week.

USS Helena, *one of the first ships I served on after joining the US Navy.*

USS Cony, *the destroyer on which I served during the Korean War.*

After graduation from IIT, I owed the navy three years of service. I circumnavigated the globe on a destroyer during the Korean War. My ship, USS *Cony*, cruised up and down the coast of North Korea to disrupt the flow of munitions among the North's troops. During the day we would use our five-inch guns to blow up railroad tracks only to wake up the next morning to discover that they had been fully repaired during the night.

I found few things to be as emboldening as commanding a modern ship of war as an officer of the deck. I earned that position after a year of intense training on the *Cony* under the critical eyes of its captain and senior officers. And I was pretty good at it. I had studied the characteristics of the three-hundred-ton *Cony*, including its systems and response times. I knew how my commands affected the power its thousand-horsepower turbines delivered to the propellers and how its rudder forced the enormous mass to turn. It took a crew of three hundred sailors to run the *Cony*, but when I stood on the bridge, it was as though the ship and I were one; I could will it to perform exactly as I wished. I built my self-confidence on that ship. And I discovered some of my limitations, including that, unlike most people, I am unable to easily distinguish left from right.

During one exercise we were tasked, along with a half dozen other destroyers, with surrounding and protecting an aircraft carrier from submarine attacks. My job was to keep the ship a fixed distance alongside the carrier as we steamed through the Yellow Sea. This would have been easy if the carrier maintained a steady course. However, to avoid imaginary submarines, the carrier would zigzag continuously, requiring me to get my ship back alongside. Figuring out the maneuver to

Ensign Martin Cooper, US Navy, 1952.

this required solving a simple geometric problem that I had practiced repeatedly—mentally, at least. That was easy. Now I had to give orders to the helmsman, who steered the ship, and to the engine room. I would say, "All ahead full." That took care of the engine room. Then I would shake my writing hand, reminding me that it was my *right* hand, and give the order "Right full rudder!" I had to repeat that process for each of the many turns.

Thankfully, no one ever noticed the delay while I was figuring out what order to give the helmsman.

While I enjoyed being on the *Cony*, successful naval careerists were aircraft pilots and submarine officers. Realizing that the inability to know right from left would be a severe handicap in a plane moving five hundred miles per hour, I chose to become a submariner and was accepted to attend the navy's submarine school in New London, Connecticut. Our education was a combination of classroom study and hands-on exercises. In a simulated torpedo room, we were taught the procedure followed by a sailor firing a torpedo. The procedure entailed operating a sequence of valves, buttons, and levers.

At the first drill, a group of us crammed into the room. As we waited to hear our names called at random from a list, I watched my classmates one after the other try to remember and execute the labyrinthine process of throwing levers and pushing buttons in the precise order needed to fire a torpedo. As they went through the steps, they were required to narrate each action. One classmate botched the drill by missing a step, the next completed a step out of order, while the third forgot to announce each step along the way. The already tense atmosphere among we submarine officers in training grew increasingly taut.

"Cooper, this is an emergency!" I heard the training officer shout. "Fire a torpedo at once—forget about the announcements."

Like an automaton, I sprang into action without thinking. Thinking would have slowed me down. Instead, my hands were a blur as I threw levers and adjusted valves with precision, maximum efficiency, and in the right order. Like a trio of exclamation marks to punctuate the drill, I slammed the breech door shut, locked it, and pounded the firing button. I watched the instructor make a notation on his clipboard. "Well done, Cooper," he said. I couldn't even remember what I'd done.

Competitive pressure helped me that day and throughout the training so that even though I did not finish sub school as *the* top student among a class of eighty-nine officers, my "excellent performance" during torpedo drills was noted and I was one of the top four submarine officers.

This was no idle honor, as it gave me the right to pick the locale of my first assignment upon graduation in 1952. Hawaii sounded wonderful, so I picked USS *Tang*, one of the last of the non-nuclear submarines, stationed at Pearl Harbor. Over the next eighteen months, I became both an expert scuba diver and an expert Waikiki Beach bar cruiser.

I found submarine duty to be agreeable, especially for officers. We spent weeks at sea conducting readiness exercises and then weeks ashore maintaining the sub.

USS Tang, *the submarine I chose to serve on after submarine officer training school.*

I had no claustrophobic aversion to the compactness of submarine living. I was seriously considering a naval career. The navy required me to learn how to think fast, make quick decisions, and focus on getting things done. Training focused on practicing for emergencies, putting us in confusing situations and assessing how we solved the immediate problem. This training in quick ingenuity served me well after I left the navy.

Because I ultimately decided to leave. The bars and hotel shows in Honolulu eventually became tiresome. I also missed Harriet, my girlfriend back in Chicago, whom I'd been dating since college. I sat down and made a list of pros and cons, agonizing over the decision for weeks. At last, I decided to resign and return to civilian life, especially to marry Harriet.

The day after I submitted my resignation, I received a Dear John letter from Harriet. We had been corresponding during my year and a half on the submarine, but she was done waiting.

Twenty years later, I ran into Harriet. It turned out that her parents had written the letter, not Harriet. They thought I was too old for her and would make an unsuitable husband for their daughter. So much for my pro-con list.

FROM TELEPRINTERS
TO TRANSISTORS

After a brief wallow in the sadness of losing Harriet, I re-engaged with my pre-navy dream of being an engineer and inventing things that would change the world. I would have loved to work for Bell Laboratories, the large-scale research arm of AT&T.[1] At the time, it was the leader in creative commercial research. But it was only interested in people with advanced degrees, which I didn't have. Later on, AT&T would have a profound effect on my career as a business customer—and antagonist.

My second choice, the Armour Research Foundation, was a division of my alma mater, Illinois Institute of Technology. It turned me down. It was a source of some satisfaction when, years later, I became a trustee of IIT, which owned Armour Research.

So, I accepted a job with the Teletype Corporation in Skokie, just outside Chicago.

In the 1950s, Teletype was the dominant manufacturer of teleprinters. A teleprinter was something like a cross between an electric typewriter and an early version of email—it could be used to send typed messages over telegraph wires or phone lines. Teleprinters aren't much in use anymore, although you can still find them in places like airports under the "TTY" sign. Those, however, are electronic teleprinters. When I worked at Teletype, the teleprinters were, except for the motor that drove them, completely mechanical.

17

I was hired as a research engineer, but my first assignment at the company was in the test laboratory, where racks of Teletype teleprinters clacked away for twenty-four hours a day. If a printer failed, I disassembled it to find the source of the failure—they were mechanical, after all. I replaced the worn parts and reassembled the printers. For some reason, each time I reassembled a printer, I had one or two parts left over. That didn't seem to matter much: the printers still worked. I once did the same thing to my Triumph Stag. Perhaps for that reason, after six weeks, I was prematurely elevated to the position of electronics researcher.

This promotion reflected a dawning realization among Teletype's management. They knew their leadership in mechanical printers was solid but recognized that electronic devices might have a role in the company's future. So they created the electronics research department, in which I worked. I assumed, naively, that this was my chance to help create a new electronic future for the company. Management's talk proved to be merely lip service. The new electronics research team consisted of a few engineers dabbling on the fringes of Teletype's vast mechanical printer business.

My supervisor, Cliff Seidler, tasked me with developing an innovative way to package a new device called a Teletype multiplexer. This device combined signals from several Teletype machines, transmitted them over one wire to a distant location, then distributed them to multiple Teletype machines on the other end. The initial multiplexer was fabricated on a metal chassis with parts wired together by soldered connections.

To help push Teletype into the electronic future, I designed a printed circuit board for the multiplexer. This was a plastic board on which etched copper wires connected the electrical parts mounted in its holes. A printed circuit board would be more reliable, cheaper, and less prone to manufacturing error than the metal version. After laboriously laying out the pattern for the circuit, I ran into a problem. The company I selected to manufacture it failed to produce a single useful board, despite their optimistic promises. My carefully designed circuit board went nowhere. Today, virtually all modern electronic devices are built using some form of printed circuit. Though it was never produced, the one I designed for Teletype was one of the first printed circuit boards designed for civilian use.

Teletype was owned by Western Electric, which was in turn a subsidiary of AT&T. At the time, AT&T was the dominant communications company in the world. In fact, it was the most dominant company of any kind in the world. So, in a sense, I did work for AT&T, as I had hoped after leaving the navy. Yet working for Teletype was not the same as working for Bell Labs, which had been my goal.

Bell Labs was the crown jewel in the AT&T empire, pioneering basic and applied research in many fields. Teletype sat low in the corporate hierarchy and suffered from shortsightedness. Its leaders could sense the electronic future but couldn't figure out how to take the company there.

I now knew, from my brief experience with electronics, that I wanted to be an electrical engineer. But my budding career had hit a wall at Teletype. I was working in a company that made great mechanical products but had no capacity to shift into electronics. When a recruiter invited me to interview for a job in Motorola's Applied Research Department, I accepted. What could I lose?

On a windy November day in 1954, I found myself sweating at a chalkboard in the Motorola headquarters building on Augusta Boulevard in Chicago. My board of inquisition was two Motorola executives, both PhDs. Bill Firestone was the vice president of research and development, and Jona Cohn headed the Applied Research Department. They asked me to derive the Nyquist–Shannon theorem for sampling an analog signal.

I had no idea who Nyquist and Shannon were and only a vague idea of what sampling was. But Bill and Jona patiently led me through the process and gave me my first lesson in communications theory. They were, indeed, testing me, but they were also trying to impress me with the intellectual challenge of work at Motorola.

They succeeded. After less than a year at Teletype, I accepted Motorola's offer to be a senior development engineer in Applied Research at $115 per week (about $1,140 in today's dollars)—this was a bit more than my $97-a-week Teletype salary. But I wasn't taking the job for the money; I just wanted out of Teletype.[2] Motorola also paid me a token one dollar in cash for the legal rights to any inventions I might create during my employment. This apparently egregiously low payment ended up being the best deal I ever made. I would enjoy the privilege of working alongside an extraordinary group of dedicated professionals with superb management and strong market position.

Transitioning from Teletype to Motorola was an experience in culture shock. At Teletype, I sat in a room with more than one hundred engineers. Lunch and quitting time were triggered—and followed religiously—by a bell. At Motorola, just a few of us shared a small office adjacent to our lab. Discussions underway in the late afternoon continued until they were over, the clock be damned. I rarely got home before seven or eight in the evening, energized and happy. Barbara Shaner, whom I married in 1955, tirelessly supported my zeal for work. That same year, I started a master's program in electrical engineering and mathematics (in night classes) at IIT, which I completed in 1957. I then became an instructor there for a few years.

The Applied Research Department of Motorola was located on the second floor of the headquarters building, an ancient structure that was falling apart around us. On hot summer days, drops of roofing tar leaked onto our heads. It was not unlike the tenement classrooms at IIT. And, like IIT, it was the perfect learning environment. My teachers were brilliant: Ed Bedrosian, Carrol Lindholm, and Bill Cannon were experienced researchers. I reported to Harry Kotecki, a skilled circuit designer.

Within a year I felt as if I carried my weight, even if I was still somewhat inexperienced and overconfident. I also came to realize how much professional success depends on not only skill but also luck.

One of my early assignments was on a project for the federal government building cryptographic equipment used for sending secure messages. My role, unsupervised by anyone, was to design new circuits, and I had to come up with a new testing device for the equipment. The normal process was to make a hand-built "breadboard" for prototypes to be sure the device worked. Only after the breadboard was tested thoroughly was a production version created.

In my youthful arrogance, I designed the device and, without testing it, requested that the model shop fabricate a finished chassis. I was elated. I had taken this assignment, part of an important government project, from start to finish all on my own.

But the fabricated version from the model shop didn't work. My decision to skip the testing step had backfired. In a panic, I rushed to figure out what I had done wrong, redesigned the circuit, tested it, and by pure luck was able to reassemble the parts on the fabricated chassis.

After my relief and euphoria faded, I realized how stupid I had been to ignore the important step of testing, and how lucky I had been that my panicked redesign worked. The process of creating and testing a breadboard before taking it to production existed for a reason. As I reflected more on the episode, I looked at it another way. I had taken a risk and was able to muddle through without dramatic consequences (perhaps it was my quick ingenuity from the navy at work). What I initially saw as luck was actually the ability to confront a setback and figure out a different way forward. I would count on this ability (or "luck") for the rest of my career.

My best luck was joining Motorola at an incredibly exciting time in the company's history. Motorola had been founded in 1928, the same year as Teletype. Yet at its quarter-century mark, when I joined, the company was in a far different position than Teletype. Motorola was at the frontier of electronics research and commercial products, especially when it came to what has been called "the 20th century's supreme electronic invention."[3] This was the transistor.

The transistor had been invented in 1947, only seven years before I joined Motorola. It was originally created by researchers at Bell Labs in New Jersey. Up until this point, electronic devices such as televisions and early computers relied on vacuum tubes. These were pretty much exactly what they sound like: glass tubes that enclose a vacuum. They conducted electricity and, in the 1940s and 1950s, represented the leading edge of electronics. But they were a "technological dead end," too large and generating too much heat to be of continued use in electronic devices that kept getting smaller.[4]

The transistor was something different altogether: smaller, more efficient, and better-performing than vacuum tubes. Since its invention, several companies had been experimenting with different uses for the transistor. Motorola made transistors using germanium, a crystalline material. Germanium was later replaced by silicon. The company's new Semiconductor Division, located in Phoenix, was a leader in transistor research.

At the Applied Research Department in Chicago, we became a principal internal customer of the Semiconductor Division. In the midst of working on the cryptographic machine for the federal government, we discovered that one of the circuits was unreliable. This was a complex machine and we were otherwise ready to ship it; there was no way we could redesign it and still meet the delivery schedule. Only a few weeks before, experts in an adjoining lab had taught me how transistors worked, and I wondered if, because of their small size, they could solve my challenge here.

The Semiconductor Division had recently shipped us some transistor prototypes. The cutting-edge electronic parts arrived in a wooden keg that looked like a miniature wine barrel. Inside was a jumble of transistors with their leads tangled. Other engineers had already absconded with the best-performing ones. I dug through the pile and found a few that I hoped would work. With some creative soldering, I hand-wired them into the existing circuits inside the cryptographic machine. To my great surprise (and relief), they worked, and we shipped the product on time.

This cryptographic machine turned out to be the very first Motorola product that used transistors.[5] That's how innovation feels in real time. We like to romanticize it and imagine supergeniuses in the comfortable surrounds of a well-appointed laboratory waiting for lightning to strike. In reality, innovation is messy.

It happens inside buildings with leaky roofs. It happens under the pressure of customer timelines. And it happens when you're trying to solve problems—even those created by yourself.

◌⚬◌

In 1956, after not quite two years with Motorola, I was promoted and put in charge of my own research group within the Applied Research Department. This meant I had some freedom to choose the projects I worked on, rather than only being assigned them from above. I was hooked on transistors and eager to figure out how to use them in other creative ways. One of the places I looked was the car phone.

I had not yet worked on telephony. Telephones had been around for about eighty years at that point, and the national telephone network was the Bell System. The company had a monopoly, approved and regulated by the federal government, over every facet of the wired telephone system. That infrastructure—well established, monopolized, with little room for new ideas—held no attraction for me. What did draw my interest was the still-emerging field of *mobile* telephony—that is, phone calls made wirelessly rather than through the fixed wires that AT&T dominated. The only way to communicate without wires was to send signals over the radio waves, and this was something at which Motorola excelled.

In its nearly three decades of existence, Motorola had made its name, literally, from the design and manufacture of car radios. The Galvin Manufacturing Company, the first iteration of the firm, had given its first car radio the brand name "Motorola" in 1930. Soon after, the company began selling one- and two-way radios to police departments across the country and would eventually make two-way communications dispatch systems for taxicab companies. (Collectively, these private two-way radio communications systems are known as "land mobile" services. They continue to be essential to many industries and companies today.)

I was intrigued by the challenge of extending the ability of people and organizations to communicate independently of wires. The importance of that ability had been demonstrated during my naval service on ships and subs—use of wires wasn't an option on those, and radio communication was essential to safety and security as the vessels moved around the world. Even before that, my childhood fantasizing about undersea living and long-distance tunnels depended on the ability to communicate without wires.

The notion of mobile telephony had, by the late 1950s, been around for some time, and car phones represented the frontier of this area. It didn't fall under the Bell monopoly, but AT&T was working on the technology. Their Mobile Telephone

Service (MTS) had debuted in the 1940s even before researchers at Bell Labs had first proposed the cellular concept. (More on that in a later chapter.) The equipment for Bell's mobile phone system was made by Motorola.

To even use the label "car phone" is an overstatement. It was little more than a two-way radio mounted permanently on the dashboard of an automobile—but this meant it was right in Motorola's wheelhouse. I saw a chance to bring together my interest in wireless communications and my desire to use transistors in a wider range of devices.

The two-way radio that constituted the car phone was equipped with a device called a decoder-selector. A landline caller would call a special operator, who in turn would dial the phone number of the car phone. The system relied on the

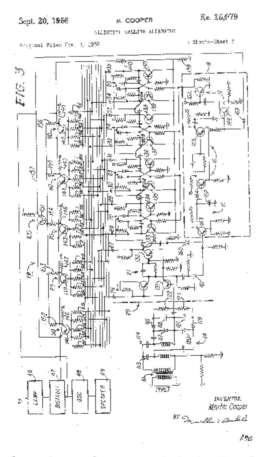

The schematic of my application of transistors to the decoder-selector for early car phones.

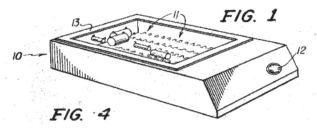

Prototype of new design for the decoder-selector.

decoder-selector to count groups of pulses sent by the operator. Each decoder-selector responded to a different combination of pulses. If the right combination was received, a bell rang in the car, and the person in the car answered the call.

Yet decoder-selectors were unreliable. They relied on a Rube Goldberg assembly of mechanical parts that looked like a large, primitive watch. As a mental exercise, I decided to design a transistorized decoder-selector that did exactly what the mechanical one did but electronically—with no moving parts. The challenge was to minimize the number of transistors used, because in those early days, they were expensive. Backward estimates of transistor costs vary. One Motorola engineer recalled that, in the 1950s, a single transistor cost about fifty dollars, or $432 in today's dollars.[6] Overall, transistors have experienced a "staggering reduction" in cost, "roughly a factor of a billion over the last 50 years."[7] At the same time, they've shrunk in size and improved in performance by comparable multiples.

My new design of the decoder-selector used thirty-nine transistors. This was a significant advance at the time, although today's smartphones may contain over two billion transistors. For the technically oriented reader, the nearby illustration shows the schematic as copied from the US patent application.

I built a model of the decoder-selector and packaged it in a box with a transparent cover to show off my achievement. Our work in Applied Research was to investigate ideas, not create products, but to my surprise, our sales staff liked it enough to offer it for sale to Western Electric, the Bell subsidiary that manufactured the mechanical version. They weren't interested. As I came to learn in future encounters with the Bell monopoly, if it wasn't their idea, it couldn't be good.

In the late 1950s, engineering graduates—especially those who wanted to work in electronics—were in short supply, and Motorola was expanding rapidly. Every department was responsible for hiring their new engineers, so they competed not only with other electronic companies but also with other Motorola departments. I found that I had a talent for hiring, bringing in as many new research engineers as the Applied Research Department could afford. This attracted the attention of the higher-ups in the Mobile, Portable, and Two-Way Radio Department, which was the heart of Motorola. Evidently, I was hiring five research engineers for every one product engineer they were able to hire. As I was about to find out, product engineers were far more important to the company's profitability than research engineers. The latter were important, but our work in Applied Research was always one or two degrees removed from direct customer interaction. Those in charge of the customer-facing business wanted more product engineers.

In 1958, Jona Cohn, who was then my boss, called me into his office to tell me that the chief engineer for mobile products was considering making me an offer. Jona preferred that I stay with him in Applied Research but told me the decision was mine. An offer had also been made to my colleague Roy Richardson, who declined it. Despite being second fiddle, I listened to the pitch, which came from John Mitchell, assistant chief engineer for mobile products. His pitch was more or less as follows: "Research may be interesting, but if you want to be in the mainstream, you need to build products people will pay for. We're the guys who earn the profits of the company—this is where the action is. Besides, we have all the fun."

John invited me into one of his laboratories, where engineers were designing the receivers and transmitters at the heart of two-way radios.

"We do research here, just like you've been doing. But our research is going into products that will ship," he said. "Just look at these front-end tuned circuits." I saw a stack of metal plates with adjusting screws at their ends.

"We're making a resonant cavity out of sheet metal that will make our radios perform better, that will let more people fit into the limited number of radio channels available to us," he continued. "This is not theoretical research. These devices need to be manufacturable, low cost, and consistent in performance. Sure, we understand the theory, but our tools are common sense and a pair of tin snips."

This, I learned, was the real Motorola way. Research was respected and supported, but commercial viability—usefulness by customers—was paramount. This was how Motorola, despite being the same age as Teletype and much smaller than AT&T (with far fewer resources than Bell Labs), was able to remain innovative and stay ahead of the competition. Customer use and value were our lodestars. (This concept is repackaged today in Silicon Valley as "customer development" and "lean startup.") The Bell System, as a monopoly, saw itself as able to dictate to customers rather than cater to their needs.

My recollection of Motorola's ethos is echoed by others who worked there in the 1950s and 1960s. Engineers in the Semiconductor Division recall that it was the sales forces who really helped push Motorola's work on transistors and the eventual move into silicon. Automobile, radio, and television customers were asking for more powerful and reliable devices. It fell to the engineers to figure out how to meet their technical demands without breaking the bank.[8] Innovation happens at the intersection of invention and utility.

Now, here was John Mitchell offering me a chance to move to the center of innovation at Motorola, in mobile products. He stripped away all the distracting jargon and complexity from technology, as he did with every engineering or business problem. His method was about understanding the underlying principles, simplifying the problem, and solving it.

John was laying down a challenge. This was the big-time; these were the pros. I would no longer be fiddling around with research ideas that only sometimes turned into products. If I could rise to the occasion, I was the guy he wanted.

Challenge and no-nonsense problem solving—this pattern would be the basis of my professional life for the next fifty-five years. I had my doubts about the position but was ready to accept it. No one had told me what my salary would be. It had taken me four years to get my salary up to over $10,000, the equivalent of nearly $100,000 today. When I'd left the navy, my definition of ultimate success was earning $10,000 a year.

My next stop was the office of Bob Peth, the chief engineer for mobile products. "I'm seriously considering your offer," I said, desperately trying to disguise my eagerness to accept John's challenge, "but I was hoping to get a significant increase in salary."

"We were thinking about $12,200," Bob responded.

I accepted immediately, and we shook hands. I was off in a new direction, as an engineering section manager.

John Mitchell was my mentor for the next twenty-five years. He welcomed me, introduced me to my team and colleagues, and sat me down, as he did many times thereafter, to set down the rules. Rule number one was to make sure that Frank Pawlowski never set foot in his office or in any other management office.

Frank was a superb engineer whose strength was not necessarily his technical skill but rather a dogged persistence and uncompromising standard of excellence. His expertise was in receiver design. Frank also had an intense need for appreciation. He was constantly coming to management to either demonstrate his latest achievement or complain about the obstacles to even greater achievement.

Ultimately, Frank taught me the intricacies of receiver design and, unwittingly, the essence of managing difficult people. Within hours of my new assignment, Frank had me in his laboratory adjusting an FM receiver discriminator to show me how he had achieved an unprecedented linearity. Since I was a digital engineer, I barely knew what a discriminator was and even the concept of linearity was new to me. I managed to learn enough to intelligently work alongside Frank, but to fulfill my promise to John, a much more involved relationship became necessary.

In the ensuing weeks, I found myself a part of Frank's social life as well. We started with a weekly game of ping-pong at the home of Roy Richardson. Frank played ping-pong like he did everything else in life—as though winning was life or death. My ping-pong skills improved considerably.

I discovered that Frank had been a horseman in the Polish army. I had always wanted to learn how to ride a horse. We decided to take riding lessons together. This soon evolved into Sunday morning rides through the spectacular forest preserves that surround Chicago. These rides were an integral part of my life for the next two decades. In 1962, a few years after my first riding lesson, in partnership with my friend Lee Cannon, we bought a beautiful seven-year-old Palomino horse for the grand sum of $700. His name was Cash and Carry; we called him Cash. He had suffered abuse at the hands of his previous owners, an unscrupulous pair of brothers. Suffice it to say that one of them killed the other within a few years after we bought Cash. He was a spirited horse whose headstrong temperament matched my own. When we connected, we moved as a single biological entity. That horse made me feel beautiful.

During an exercise session on his last day, after twenty years together, he playfully jumped over three fences set up in the corral, kicked his heels in the air like a young colt, dropped to his knees and fell over, dead. Cash was as strong, as energetic, and as willful when he died at the age of twenty-seven (his nineties in human terms) as he was when I bought him at the age of seven. He was with me through so much of the work to develop the cell phone. But I'm getting ahead of myself; I haven't even moved into portable products yet.

CAR PHONES
Encountering the AT&T Monopoly

I n Woody Allen's 1972 movie, *Play It Again, Sam*, Tony Roberts plays the main character's work-obsessed friend, Dick Christie. As Christie moves about his hectic life, in apartments, restaurants, and elsewhere, he repeatedly calls his office to give instructions on how he can be reached.

"I'm at the Hong Phat Noodle Company. It's, uh, 824–7996."

"I'm no longer at 431–5997. I'll be at Mr. Felix's . . ."

That's how we did it before cell phones. Christie's imprisonment by the fixed telephone wire is almost as real as if it was tied around his waist. If the movie had been made thirty years later, he would have been freed from the wire, in constant touch, fielding calls and exchanging text messages.

But why didn't Christie have a handheld cell phone? Why didn't a modern cellular system exist in 1972 for making calls from anywhere? Conceptual designs for an automobile-based cellular system had been created in 1947 at Bell Labs, a quarter-century earlier. By the late 1960s, primitive and experimental vehicular systems existed in many parts of the country.

A handheld phone was not practical using existing batteries; their poor capacity couldn't support long phone calls. And there weren't enough radio channels in any city to provide decent service for more than one hundred subscribers. The Federal Communications Commission (FCC), which had authority to dedicate more radio channels for mobile phones, was reluctant to do that. A single phone conversation would tie up a radio channel for a fifty-mile radius, which meant that there were not enough channels in the entire usable radio frequency spectrum to support even minimal market demand. Most of the spectrum that would be needed for such a service was already occupied by TV stations. As we'll see, the television broadcast industry was a powerful lobby in protecting the spectrum rights of those stations.

Progress in personal communications accelerated in the 1960s based on a rocky but effective relationship between two companies: Motorola and AT&T. Today's mobile phone landscape grew out of a simultaneously cooperative and competitive relationship between them that drove the direction of technology. The story of that relationship demonstrates how our existing systems evolved and may offer clues as to how they will continue to evolve in the future. It was my good fortune to be an active participant at the center of this drama. That experience shaped how I came to see technology and its progress.

Technological development isn't shaped solely by radical breakthroughs and technical tweaks. Academic research and industrial research and development are important. But development of a breakthrough technology is determined by the marketplace, by consumers, and by the perspectives and decisions of the companies that must develop the products and services that people buy.

The development of the mobile phone reflected all these interacting dynamics. Motorola and AT&T weren't just different companies. They represented totally different views of the future of communications technology and how people could, or should, use it. These contrasting outlooks were born of their experiences with different technologies, different types of customers, and different organizational cultures.

In AT&T's technological future, Dick Christie would have enjoyed the benefit of mobile phones—but they would have been exclusively vehicle-based and limited in range. In the technological future we envisioned at Motorola, Dick Christie would get something altogether different. To get there, though, required a long road of cooperation and competition with AT&T. That road started in New Jersey, wound through Chicago, and ended in Washington.

<hr/>

In the fall of 1959, I drove through the New Jersey suburbs to a new Bell Laboratories research facility in Holmdel. The sky was as gray and featureless as the terrain. Looming impressively at the entrance to the property was a sixty-foot, three-legged water tower. It was shaped like the transistor that Bell Labs had in-

The transistor-shaped water tower at the former Bell Labs facility in Holmdel, NJ.

vented a dozen years earlier and that companies like Motorola were in the process of exploring and applying commercially.

The featureless glass exterior of the research building did little to improve the barrenness of the site. The new Holmdel facility was the first of its kind to be clad solely in glass and had only recently been occupied. Eventually, it would house over six thousand research engineers and scientists. The campus was a symbol of the technological power of the largest company in the world.

I was at Holmdel to learn more about a rumor picked up by Motorola's sales force. AT&T was reportedly designing a system to replace its Mobile Telephone Service (MTS) that served around thirty thousand subscribers in the United States. This system had been operational since 1946, when AT&T first deployed it in St. Louis. Using radiotelephones in the trunks of cars and trucks, MTS was only operational from vehicles. To place a call, you would turn a radio knob to find clear frequency, push a talk button, and call an operator. The operator would connect you to the number you wanted to call. It was known as "half-duplex," like a walkie-talkie, meaning a caller could either talk or listen, but not both at the same time.[1] The system served its subscribers so poorly that a person who wanted to make a phone call during a busy time of day had only one chance in twenty of completing the call.

A Motorola-made MTS system for car trunks, shown here in 1946.

Motorola had been making most of the equipment for the MTS radiotelephone system for fifteen years and, if it were true that AT&T was designing a replacement, I had to get us in on the action.

As part of my duties as an engineering section manager, I had more interaction with Motorola customers like AT&T. Despite AT&T's overall size, the MTS system was small and represented only a modest revenue stream for Motorola. Our bread and butter remained in land mobile radio, especially two-way public safety and dispatch systems. Still, a replacement for MTS was likely to be a much bigger system requiring more technical advancements.

The Holmdel water tower and building facade were designed to be intimidating, but I wasn't having it. I was on a mission and wouldn't be distracted. On an earlier visit to Bell Labs' Murray Hill facility, I had realized it had been a blessing to be rejected by the Labs earlier in my career. An exasperating lunch spent listening to Labs employees ponder the properties of various steel blades had shown me that I was more like my mother than I thought. I wanted to make useful products to help solve customer problems—I wanted to sell, not sit around the lunch table and muse.

Remembering my mother's ease and confidence in talking to potential customers, I approached Holmdel determined to get Motorola a piece of the new MTS.

Bell did indeed have a new concept that they called the Interim Mobile Telephone System (IMTS). It was a preliminary phase of what was to become the Advanced Mobile Phone System (AMPS), one of the first versions of modern cellular telephony. This represented a grand Bell Labs vision for reproducing the wired telephone network, but over radio instead of wire. As the "interim" label implied, IMTS was the first phase in testing and rolling out this vision.

Somebody in the Bell System must have figured out that naming a product "interim" was amateurish marketing. Who would buy a product knowing that a superior version was already in the offing? They renamed it the Improved Mobile Telephone System. The change wasn't terribly relevant, since the system was already called, simply, IMTS.

My goal was to get Motorola invited to participate in the IMTS project. I was prepared to argue to Bell Labs' management that we knew more about two-way radios and mobile telephones than anyone in the world. In fact, I was prepared to make the case that AT&T *needed* to work with us if they were to ever produce a successful mobile telephone.

At Holmdel, I left my rental car in the largely unoccupied parking lot. In the lobby, an atrium with lush plants contrasted sharply with the barren landscape outside. The receptionist made a phone call. Minutes later, Roger Cormier arrived at the reception desk to escort me inside. Over time, we would become colleagues, friends, and occasionally antagonists. I was there as a potential collaborator on an ambitious new mobile telephone system, but he never hesitated to remind me that this was a Bell Labs project and Bell Labs made the rules.

To say any Bell Labs engineer was smart is an understatement, but Roger was more than smart. He was brilliant, detail-oriented, and forgot nothing. Yet he was never overbearing. He managed to be intellectually superior without being arrogant and sharp without being cutting.[2]

On that first day, we took the elevator to the top floor, where Claude Davis waited for us. Claude was Roger's boss and the antithesis of Roger's nerdiness (no offense intended; I was a nerd myself). Claude's pale-blue eyes enhanced his warm smile, and he had a matching sense of humor. He exuded self-confidence and moved with athletic precision.

"This building was designed as a research facility," Claude said as he escorted us to a conference room. "There are no offices with windows. The periphery of the

The imposing glass exterior of the Bell Labs Holmdel research facility, designed by Eero Saarinen.

building is all aisles and the modestly sized offices open into interior aisles. A typical office, even for someone like the legendary Claude Shannon, only has a desk and chair. We don't waste time and energy on ego," he said. The idea that a research god like Shannon sat in an office the same size as a lowly engineer was astounding. At most other companies, he would have had a fancy office comparable to that of a company officer.[3] After all, he had helped prove what became the Nyquist–Shannon theorem, which had taxed my mathematical abilities in my first interview with Motorola.

We arrived at a conference room, where Claude Davis spelled out Bell Labs' plans for IMTS.

"IMTS will be an entirely new mobile telephone system. We're not going to adapt existing mobile equipment, as we did in the past. We're starting from scratch and making a purpose-designed system. But we're taking an unusual approach, at least for Bell Labs. Instead of designing the equipment for this system and having our sister company, Western Electric, build it, we're going to select outside companies to design and manufacture the hardware," he told us.

"We'd like to be part of that," I interjected, trying to suppress my eagerness. "We already build most of the existing mobile telephones." An MTS mobile telephone at that time was nothing more than a two-way radio equipped with a decoder-selector and connected manually to the telephone network by operators. Motorola was the industry leader in two-way radios.

At that moment, an older man joined us. He was dressed in coat and tie, in contrast to Claude's rolled-up shirtsleeves.

"Meet my boss, Dr. King Edward Gould. Dr. Gould is a director here at the Labs." I thought Claude was pulling my leg; nobody could possibly go through life referring to himself as King *anything*.

True to his name, King wasted no time in outlining Motorola's insignificance. "I'm aware that Motorola has expertise in the two-way radio business," he said. "But IMTS is a telephone business. It's the supervisory unit that will distinguish this new service."

The supervisory unit that King referred to was an electronic device that could make the telephone connection automatic, dispensing with the need for an operator. This device was similar in function to the decoder-selector I invented in 1955, but much more complex. It would allow a user to make and receive calls wirelessly but automatically, simulating a wired telephone.

His emphasis on *telephone* and *supervisory* shocked me. These were archaic terms, wired telephone terminology, to those of us who believed in mobility and portability, those who believed—as I and my Motorola colleagues did—that enhanced radio was the future of communications.

"The supervisory unit is the brains, the logic in a mobile telephone," he declared. "The supervisory logic is all that is new in the IMTS system. The radio is incidental. Relatively unimportant. We won't have 'the tail wagging the dog.'"

"I agree that automatic calling is important," I tried to interject. "But it's still the voice communications that our customers will experience."

"Let me stop you there," King said dismissively. "There's no need for you to instruct us about your position in the two-way radio market. We're aware, but it doesn't really matter for IMTS. The supervisory unit that Secode Corporation has built comprises the brains of our system. Of course, we'll upgrade Secode's design to meet Bell standards. Our selection of provider for the radio contract doesn't much matter."

My impulse to argue with King was squelched by a subtle head shake from Claude; this was neither the time nor the audience for argument. None of this made sense to me, certainly not from the perspective of the customer or user. Radio performance was absolutely critical to the user. Mobile phones made calls over radio waves, and radio networks were tricky, different from wired networks in many ways. Motorola's engineers were skilled at making radio networks that sounded clearer and better than wired networks.

It also didn't make sense in terms of equipment. Adding a supervisory unit would mean an additional box in the trunk of a car—in addition to the two-way radio equipment. The existing MTS system already weighed as much as eighty pounds. To me, this was just plain dumb. Having two different companies make these boxes was even dumber—too expensive and overly complex. It was like asking two companies to build a tunnel and expecting them to meet at exactly the right place under a river.

But at AT&T and Bell Labs, as epitomized by King Gould's terse words, this was the vision and objective. Making mobile telephone service more like the existing wired telephone system (that is, landlines) required adding hardware in a car trunk. For the Bell System, everything was seen as an adjunct to the wired system, and for mobile phones that adjunct extended only as far as the car trunk. For me and Motorola, the vision and objective were driven by a different logic— that of radio and the freedom it implied.

After King left, Claude told me that the decision had already been made. Motorola would have to be satisfied with seeking the radio part of the business and working with another company for the second box. He was going to oversee the selection process for the radio manufacturer. He said he would do that selection objectively, but that his colleagues at AT&T were not happy with Motorola and had suggested that we would be a bad choice to be a partner.

AT&T's unhappiness with Motorola was due to an unrelated regulatory situation. In 1956, an antitrust consent decree from the Justice Department had prevented the Bell System (through Western Electric) from manufacturing anything besides telephone system equipment. The decree also kept AT&T out of businesses such as land mobile (two-way radios) and the emerging computer industry. These were minor hindrances to the company: on the whole, "AT&T was pleased with the decree" because its monopoly was preserved.[4]

Three years later, however, the FCC opened the door to construction and service of microwave communications systems for private lines—these used radio waves, not fixed wires. It was a small move but created competition where little had existed. The move was strongly supported by Motorola because we were the largest manufacturer of microwave equipment besides Bell.[5] We were selling that equipment to Bell competitors. These competitors were replacing wired phone lines that had been the exclusive province of AT&T.

That support won us few friends at AT&T, and they tried to punish Motorola by reducing our manufacture of mobile phone equipment. At one time, Motorola manufactured most if not all of Bell's mobile phones for MTS. By the time I visited Holmdel, because of the legal challenge to AT&T's monopolistic behavior, Motorola's sales of mobile phones had trickled down to almost nothing. This was not a large loss for Motorola in terms of revenue, but the move made the ruthlessness of the Bell monopoly evident.

King's attitude might have been disappointing, but Claude's assurances of fairness in process gave me some comfort. There was an opening. I was persuaded that he would be fair and, if he was, I was certain we would be invited to participate in the IMTS program.

I said my goodbyes and headed back to the parking lot. I reminded myself that Bell Labs was the most competent and accomplished communications research organization in history. My task was to persuade Bell management that their view of Motorola's capability was simply wrong; that we were the whole dog, not just the tail. Gulp!

⌒✳⌒

As I drove away, I watched the glass building recede in my rearview mirror. Designed by Eero Saarinen (who also designed Dulles Airport and the Gateway Arch), it was "serious and austere," a "monument to architectural presumption."[6] It was also a monument to corporate presumption. AT&T was at the height of its power and influence. The consent decree had required AT&T to divest from non-telephone businesses. But its telephone monopoly persisted and was stronger than ever. Indeed, AT&T "appeared to be indestructible. It now had the U.S. government's blessing."[7] Although I worked for a competitor, like so many engineers at the time, I still idolized their research groups and felt the faint traces of my past desire to work for them.

Yet Saarinen's design implied a different story. The reflective skin seemed to cover the building like the one-way glass panels used in interrogation rooms. One of my Motorola colleagues, Chuck Lynk, would later say that "anyone who had been in any meetings with the Bell Labs people always came away with a feeling like we were dealing with a diode." That is, an electrical device that only passes current in one direction.[8]

The Bell System was able to issue proclamations to suppliers, regulators, and customers about How Things Would Be, unhindered by consumer needs or competitive influences. Their presumed dominance couldn't last forever. Hubris had to catch up with them eventually.

AT&T later abandoned and sold the Holmdel property. The still-pristine building was considered useless by its new owners and torn down. Today, the transistor-shaped water tower still stands. Unlike many water towers, however, the Holmdel one does not carry cellular antennae on it. A fitting tribute, perhaps, to the work that did—and didn't—go on there.

Back at Motorola headquarters in Chicago, I reported on the meeting's hostile tone and newly discovered obstacles. After venting his displeasure at the evils of monopolies, John Mitchell supported my view that we needed to go after the Bell IMTS business with all the energy we could summon.

A few weeks later, AT&T contacted us and set a date for a visit to Motorola by a delegation, headed by Claude Davis, from Bell Labs and Western Electric. This same group would visit our competitors and decide whom to pick as Bell's partner in the development of the radio portion of the IMTS radiotelephone.

It was critical that Motorola win this business, despite our reservations about AT&T and despite their unremitting arrogance. The buildout of IMTS was intended to be larger than the existing mobile phone service. Even though the equipment for car phones represented a small share of Motorola's overall revenue, exclusion from this contract would likely mean being left out of future work with AT&T. Like it or not (and whether AT&T liked us or not), we had to be in on IMTS.

John Mitchell, Bill Weisz, and I met with Art Reese, our division vice president, to ask his advice about how much we should disclose in the meeting with the Bell delegation.[9] We wanted the business, of course, but were not about to give away our proprietary technology. Telecommunications was an extremely competitive business. We could not maintain our status as dominant supplier of two-way radio equipment without fiercely guarding our superior technology. Art's dark mustache bristled when he talked about his long experience as a sales executive for Motorola, during which he battled the Bell monopoly.

"We have been a thorn in AT&T's side for years," he said. "So it's very unlikely that you'll get this business. But if you don't use every technological weapon at your disposal, you won't have a chance. Either go for it with all you've got or forget it."

This was just the challenge that would stir us to action. I could hear the bugle signaling "CHARGE!"

Inspired by Art, John and I reached out to the technical powerhouses in our two-way radio departments and in my former home, the Applied Research Department. We polled them for every idea, no matter how far out. John's directive was to deliver the AT&T delegation a "dazzling demonstration" of our technology at work. It would need to be something that they simply could not refuse.

We analyzed the Bell requirement specification, extracted every unique technical challenge and potential problem, and came up with solutions to every issue.

Several different technical problems needed to be solved. For example, most two-way radios at the time still used vacuum tubes. Based on some of the pioneering work being done within Motorola, we proposed a revolutionary IMTS radio using transistors; the only vacuum tube we required was the power amplifier. Likewise, the IMTS radio had to be full-duplex, able to transmit and receive at the same time. The existing MTS radiotelephones, and Motorola's two-way radios, were half-duplex: they required you to press a button to transmit and release the button to hear the other person. We needed to make the new mobile phones more like normal telephones, where you could interrupt someone without having to press a button.

There were other issues, too. I had learned enough about radio receivers and transmitters from people like Frank Pawlowski to understand that the IMTS radio needed a receiver filter to collect the signal from the other radio, a filter to prevent the transmitter from interfering with other equipment, plus a third filter to prevent the IMTS transmitter from interfering with the IMTS receiver. Why not combine these three functions into a single device? I had no idea how to do that, but I had a hunch it could be done. Sure enough, a group in Applied Research did a superb job of creating what became known as the tri-selector. This component was an aluminum cube with resonant cavities that provided the filtering selectivity. To counteract the interference from automobile ignition systems, Motorola devised a novel receiver technique that suppressed the engine interference, thereby extending the receiver's range. We called it, naturally, a range extender.

Our demonstration included advanced piezoelectric ceramic filters, ultra-stable crystal oscillators, and other devices that had not yet seen commercial service. We were required to invent some of the devices needed to solve the technical problems, and we collected data to support their projected performance. Some of these devices and techniques were still just ideas. We were not certain they would even work, but we had confidence that, if our potential customer liked what they heard, our teams would ultimately make them work. The engineers created kernels of possible designs, some based simply on my imagination, and we presented them to the Bell Labs team as finished products, a fait accompli. The solutions were simple, beautiful engineering. As predicted, most of the solutions worked for Bell and became industry firsts.

We needed to put everything we had into this demonstration, and we did. We were selling not-yet-invented technology.

But we were confident. We believed in the ability of our people to come through, even though they themselves had doubts. With the help of Roy Richardson and Chuck Lynk, I identified the creative engineers, the expertise we needed to convince Bell, and John Mitchell made sure we got them on the team.

Years later, during the cellular development period, one of my counterparts at Bell Labs trivialized Motorola's contributions to the development of mobile phones. The key technical challenge of mobile phones, he said, had to do with computers, not radios. Yet it was clear from the IMTS experience—in response to the specifications laid down by Bell Labs—the first problems that needed to be solved had everything to do with the radio component. Without Motorola's radio contributions, AT&T wouldn't have had a mobile phone.

The AT&T visit was the first of many "John and Marty" shows. We might have been intimidated, but we were prepared and confident. John taught me the fundamentals of technology showmanship, and I loved it.[10] He had a table built with raised sides that prevented viewers from seeing the objects on the table. Like magicians, we would talk about an insurmountable technical challenge and then reach into the table and present the solution. It was magical even to me, even though I'd put much of it together.

This, I had been discovering, was how projects were created within Motorola. To get any project funded, the project leader had to persuade his or her managers that the risk involved in doing that project was worthwhile. The Motorola managers were almost all engineers and were already well versed in the tricks of showmanship. An engineer who wanted an idea to be funded had to be creative not only in originating the idea but also selling it.

Of course, engineering solutions also needed to be sold to prospective customers. The Bell Labs team who visited us were experts in both technology and manufacturing. They wouldn't be sold merely by showmanship. Using his "magician's table," John and I took the Bell Labs team through the technical challenges (such as full-duplex) and revealed Motorola's solution from behind the table's raised sides. One minute, a technical challenge was an unsolved problem; the next, Motorola had solved it and the solution was right in front of them. This was deeper than showmanship: it allowed the Bell Labs team to see the uniqueness of our approach.

It worked. Bell Labs selected Motorola to develop the radio portion of the IMTS car telephone. I was elated. But now came the hard part: we had to deliver, and that meant once again pulling together the right people within Motorola.

I assembled a team of experienced radio engineers under Burnham "Burn" Casterline, my sharpest engineering manager. We reached out to every engineering and research group in Motorola that may have had expertise we lacked. The opportunity was infectious—everyone wanted to participate in solving the IMTS technical challenges. They were too much fun to leave solely to the engineering team.

As a section manager, I was supposed to pay attention to the other products in my portfolio, but I couldn't resist putting my engineering hat on from time to time. For example, our two-way radio users typically operated with one or two channels, using a quartz crystal to tune the radio to a different radio channel. If the quartz crystal drifted off frequency, the radio signal would get noisy and fail. To correct this, our service shops would replace the quartz crystal for a particular channel and send the customer on his or her way.

Our IMTS radiotelephone worked differently. When a user lifted the handset of an IMTS telephone, the phone searched for a free radio channel on which a conversation could be held. Rather than using a quartz crystal to tune to the channel, the IMTS radio used a costly device called a channel-element that housed a crystal and the electronics that kept the radio tuned. An IMTS telephone had as many as eleven channel-elements, each of which had to be set to a precise frequency and occasionally readjusted. If the channel-element drifted off frequency, the telephone would stop working.

Further, the test equipment needed to adjust channel-elements was a precision device that cost thousands of dollars and required a skilled technician to operate. This seemed to me to be a terrible defect in the IMTS system, one that could affect the commercial success of the product. We couldn't expect every service shop to buy the test equipment and train technicians to operate it. That was a deal breaker, and this was a problem.

At the time I carpooled to work quite often. Unlike other people's preorganized systems, which were common at the time, we engineers were too independent to have a set routine and decided night by night on who would drive the next morning. During one passenger ride to work, as I daydreamed during a lull in the conversation, I conceived a simple and inexpensive test device that would accurately adjust the channel-element that then tuned the IMTS radio to its proper radio channel. I called my invention the IMTS Calibrator. It could be operated by an unskilled service person, who would only need to listen to an audio tone and watch a flashing light slow its pace of repetition as the channel element was adjusted. When the light stopped flashing, the channel element was tuned. The IMTS Calibrator would be inexpensive enough that service shops could afford one. I patented the device, assigned engineers to design it, and introduced it into the product line without a single planning meeting or marketing discussion.

We still needed to surmount the ultimate IMTS challenge: the supervisory unit. This was what King Gould was so focused on, the part of the IMTS radiotelephone that would make it work like a wired phone. The development contract for the supervisory unit was awarded to the Secode Corporation. King had called the supervisory unit the "dog" for which the radio was the tail. I'd made up my mind that Motorola could and should build the entire "dog," Secode be damned.

As the mechanical design of the IMTS radio took shape, our objective was to fit the supervisory unit into the same box as the radio. I knew that Motorola had to design and build its own unit. It didn't make sense to have two separate boxes, two devices made by different manufacturers filling up the already meager space in a car trunk. With the existing MTS system and its large equipment, the current power drain was so high that a car's headlights would dim whenever the system was used to make a call. That couldn't continue.

I envisioned what the supervisory unit circuitry would look like: a circuit board folded into book form that fit perfectly into an existing part of the radio enclosure. I lucked out again. Although the engineers who designed the chassis for the IMTS radio had no idea what the supervisory unit would look like, they inadvertently left exactly the amount of space we needed for the supervisory unit.

The Secode supervisory unit used several mechanical devices to achieve its functions. Motorola was building the radio portion of the IMTS using transistors, moving away from reliance on mechanical components. It was illogical to have a technically advanced transistorized radio and an archaic supervisory unit. It could be much more reliable if designed with transistors and no mechanical devices. I had a rapport with Dick Adlhoch, a brilliant young engineer who shared my minority position as a digital engineer, a rarity in Motorola's radio frequency engineering environment.

"Dick, we need to design a digital interface to our IMTS radio that controls all the signaling and channel selection," I said.

"By 'we,' I assume you mean me, and I'm your man," he responded.

"There's more to it. You know that transistors are selling for over two dollars. We're competing with cheap relays, so you have to be clever and do this with fewer than one hundred transistors."

The next day Dick returned after studying the specs. "It can be done, but I'll need at least 120 transistors and a couple of relays."

"Let's negotiate," I told him. "One hundred and nine transistors and one relay, and a bonus if you get rid of the relay."

A month later, we had a design that needed only ninety-nine transistors and zero relays. The Motorola IMTS unit contained the entire radio and supervisory unit in a single metal box. This supplanted what Secode was making: few if any of their supervisory units were ever sold. Bell Labs paid Secode for their initial design of the IMTS system, but their plan to market a separate supervisory unit—and one that was still mechanical—was wrong for customers and users.

Not all my ideas about the IMTS radiotelephone were useful. Despite Motorola's ability to make the supervisory unit digital with transistors, another key component, the power amplifier, still relied on vacuum tubes. The amplifier produced the output power of the telephone. It was not possible, in 1962, to produce the twenty watts of power specified by Bell Labs with semiconductor devices. Our vacuum tube supplier, Amperex, offered to build a special new tube specifically for the IMTS radiotelephone. Having a major company design a unique device just for my radio felt like the big-time. I proudly mentioned this to Bob Peth, John Mitchell's boss.

He was not impressed: "Why take a risk on a new design when we've got a tried and true tube that we've used for years in our two-way radios?"

"I know that Bob, but that tube is overdesigned. It's capable of handling over eighty watts—four times more than we need."

"That's right! The old tube is reliable, we know it well, and it will be coasting along in your radio. Sure, it will cost a bit more, but the risk of replacing failed units also has a cost."

Deflated at using something old, but persuaded by the logic of Bob's argument, I took his advice. Ten years later I checked with Motorola's service department. There was no record of any of the old vacuum tubes *ever* failing in service. The lack of stress and the overdesign made them ultrareliable. There are IMTS radiotelephones still operating in rural areas today that are six decades old and working perfectly.

Lesson learned. The worst thing that an engineer can do is adapt technology merely for its own sake rather than for meaningfully improved user functionality.

The IMTS radiotelephone was a great success. Now, the car telephone worked just like a home or office phone. In larger cities the additional IMTS radio channels rapidly filled up, and waiting lists grew. Hundreds of thousands of Motorola radiotelephones were sold to Bell operating companies as well as independent Radio Common Carriers (RCCs). (The Bell System limited its IMTS service to forty thousand subscribers initially.) At $2,000 a pop for a radiotelephone in 1960 (the equivalent of $17,000 today), my team had a hand in starting what became a billion-dollar business. Partly as a result of my success in helping Motorola secure the IMTS contract, I was elevated to chief engineer within the Mobile Products Department.

Motorola's relationship with the Bell Labs people at Holmdel grew as each group discovered the strengths of the other. During Motorola's frequent visits, we ate together in the employee cafeteria. The Labs engineers would often skip dinner at home and bring in pizza instead.

Even King Gould warmed up. During a cafeteria lunch, he observed, "Marty, you are the personification of hypertension. Most people think hypertension is high blood pressure. But it's not—it's having a superfluity of energy ready to explode at any moment."

I didn't agree. The environment at Motorola was full of highly driven people, like my mentors Bill Weisz and John Mitchell, and later others like George Fisher. We were not hypertensive; it was the rest of the world that was slogging along at a snail's pace.

On that same visit, I spent the day with Roger Cormier and his colleagues and then headed off to my hotel at sunset. I must have been daydreaming, because I found myself driving randomly through a residential neighborhood, totally lost. Flashing headlights behind brought me back to the real world. Roger had followed me, knowing how flaky I could be, just in case I made a wrong turn into the maze of suburban streets and cul-de-sacs that made up the newly built Bell Labs neighborhood.

"Do you play Go?" Roger asked me through an open window. "No," I replied, knowing only that Go was a game played mostly in East Asia with what looked like white M&Ms on a crosshatched board. "But I'm willing to learn if you think I'm up to it." I followed him to the home of one of the engineers with whom we had been working.

Despite the encouragement of enthusiastic teachers, Go was a mystery to me and remains so today. But I was now an insider within the community of Bell Labs nerds. These relationships served us well when a new contract opportunity with AT&T arose.

This one was for the fixed radio stations that complemented the mobile telephones we were already building. It was a technology Motorola knew as well as mobile radio, although the fixed station product line was not in my department. Claude Davis called the next day.

"Marty, you guys got the mobile telephone contract because you deserved it, but it was tough getting AT&T management to buy in. I had to persuade them that no one else was remotely qualified. They agreed on the condition that I had to give the IMTS base station award to someone else," he said. That someone else would be Motorola's main competitor, General Electric.

"Claude, I appreciate the heads-up," I told him. I understood the situation but wasn't going to accept it. GE was our competitor. If GE got the contract, it would learn from the relationship with Bell Labs, and that would make it stronger. We had to beat GE. And besides, it was a matter of pride.

This meant that Motorola's next demonstration had to be beyond great. I needed to find some new technological morsels to impress the smartest engineers in the telecommunications world.

The participants were the same, the magician's table was there again, but we were able to come up with new displays of technological magnificence. The centerpiece was aimed at the most rigid of Bell Labs' specifications, known by the technical term "final amplifier linearity." One of the engineers in Roy Richardson's research department had conceived of a new way to improve the performance of the high-power amplifier, the part of the base station that delivers power to the antenna. It was theoretically possible, but one had never been built and tested. The concept was clever and, properly explained, would impress the hell out of an engineer. In the weeks before the Bell Labs visit, one researcher put the pressure on by creating an experiment that proved the principle of an ultra-linear power amplifier. It may not have been practical, and I don't know whether it was ever implemented, but that was irrelevant. Motorola understood the technology and demonstrated that we could stretch in creative ways to solve real-world problems.

Once again, our engineering skills and showmanship carried the day. Claude stuck out his already bruised neck and told AT&T management that Motorola was the only company qualified to execute the base station contract. (I had enormous respect for Claude's ability to rise above the politics and approach things rationally, not letting emotions distort the technical decisions, which is one of the reasons why several years later I persuaded him to join Motorola. The salary I offered him was greater than my own. It shocked my bosses when I told them, but they stood by me, and Claude enjoyed a successful career at Motorola.)

We had done a good job of selling our capabilities to Bell Labs and AT&T and had won development contracts for the new IMTS mobile phone and base station. Now we had to produce results. The schedule was aggressive, and Bell Labs was uncompromising in enforcing their tough specifications. For much of 1963, the year following the contract award, we endured a series of crises, seven-day weeks, and sleepless nights. What started as a Bell Labs development project for a mobile telephone and a base station was now a lot more complicated.

Claude Davis of AT&T's Bell Labs. This picture was taken after Claude joined Motorola.

It was difficult dealing with the people at Bell who had unlimited resources and high standards, as illustrated by one particular occasion. Roger Cormier from Bell Labs approached me in our lab (he was now spending more time with us than at home) and said, "I'm afraid the paint you've proposed for the base station cabinets doesn't meet our specs."

"Roger, I knew you were sensitive to that spec," I told him, "so I dealt with the cabinet builder personally." The supplier's engineer had shown me the sample painted metal square that had been tested on a machine designed to measure paint adhesion. The machine had a calibrated cone, around which the sample was bent. Contrary to the spec, the paint always cracked at the small end of the cone.

"You can't argue with the machine," I said.

"As a matter of fact, I can," he retorted. "I talked with the Labs engineer who invented the machine. He says you aren't meeting the spec, so you must be using the machine improperly." Argument over, we fixed the paint.

Meanwhile, Motorola also had to keep in mind the Radio Common Carriers, the non–Bell System providers of radiotelephone service. These were independent businesses who offered telecommunication services to the public in competition with Bell. They had effectively been created by the FCC to keep AT&T from dominating the incipient mobile telephone business. In 1949, when Bell first asked the FCC for spectrum allocation for mobile telephone service, the FCC awarded half the allocation to AT&T and half to companies that would become the RCCs. Little noticed at the time, it was really "the first FCC-created competition for the Bell System."[11]

The first Motorola IMTS radiotelephone, produced for and sold to companies other than the Bell System. The control unit (the handset and cradle) was designed by Motorola.

By the early 1960s, there were hundreds of these operators in the United States and throughout the world—collectively, they were important Motorola customers. The Bell Labs IMTS program was not due to be completed until 1964. I wanted to ship radiotelephones to RCCs much earlier. To accomplish that, Motorola would need to provide not only the mobile telephone and supervisory unit but also the control unit and handset that would reside near the dashboard of the car. This meant that we needed to either build or buy the parts of the product line that were being developed by Secode and Kellogg Switchboard and Supply Company.

A team of engineers and marketing people was assigned to visit with Kellogg to get them to build a special version of the equipment that they were building for Bell. Though Bell was trying to build the system, it could not monopolize the equipment, and from their point of view, they wanted lots of other companies to build equipment that hooked into their system (much the way apps work for the Apple platform now). The control unit, what was formerly called the telephone instrument, was the only part of the equipment the customer would see. For Bell system end users, the control unit was designed by Dreyfus Associates, the prestigious design firm whose clients included GE, Radio Corporation of America (RCA), and many others. It would, of course, have the Bell logo on it. Claude

IMTS control unit with push button and rotary dial and handset.

Davis had shown me a picture of the Dreyfus design, and I hated it. It was too big, too round, and too old-fashioned.

Gary Cannalte, a bright engineer who had no degree but was smarter than many of our traditionally educated engineers, took on design of Motorola's version of the control unit. With the support of the industrial design team, he produced a beautiful unit that was more modern and functional than the Dreyfus-designed Bell Labs unit.

Later, Bell Labs once again selected Motorola for expansion of the IMTS radiotelephone into a new frequency band, at 450 MHz. Motorola C&E, the Communications Division sales subsidiary, created a special sales team for IMTS. They were successful beyond our wildest dreams. Motorola's IMTS radiotelephone, with both manual and automatic versions, had the market to itself. The RCCs deployed them aggressively. The Bell System version appeared within eighteen months. In the early years, Motorola had almost 99 percent market share.

◦✳◦

By 1965, I had been in the Mobile Products group for seven years, first as assistant chief and then as chief engineer. Although the IMTS program was my team's most important contribution, we had also introduced the first radio-controlled traffic light system, which switched the traffic light patterns during the day via radio signal. The system required a box next to the cast-iron traffic light box. The system receiver was designed by Frank Pawlowski. For our box, we used sheet metal, which was less expensive. In a demo of the durability, one of our

Cartoon depiction of me working (and falling!) on the radio traffic control system.

salespeople tried to show that the sheet metal was just as strong as cast iron and broke his toe on a cast-iron door. The technology was sound, but we were working ad hoc.

As assistant chief engineer at the time, I helped the team install the first units. Professional installers would have used a special crane. We bought a ladder from a hardware store, which we set atop the traffic control box, so we could mount the equipment on the traffic light pole alongside the existing light system. Nobody fell, but there were close shaves. The system was deployed in Washington, DC, and Detroit.

The development of the IMTS radiotelephone was a crucial step in my career: shepherding a complex product from conception to production gave me self-confidence. And IMTS was a first taste of later battles with the Bell System.

The final Bell Labs IMTS development program, in the late 1960s, was for the new Metroliner train that ran 225 miles between New York City and Washington, DC. This system was unique in that the telephones automatically switched from one system to another—the "handoff"—as the train moved beyond the range of its base station. Claude Davis called me.

"Marty, we're giving this contract to General Electric," he said flatly. "There are only a few IMTS phones that need to be developed for the train, and they will operate in a different frequency band, so it will be a waste of your engineering effort to take this project on. Besides, I'm tired of battling AT&T management each time we award a contract to Motorola."

We shed no tears over the award of the Metroliner contract to GE. The dollar volume and number of phones on the train were insignificant. More than that, our emerging vision of truly portable mobile telephony would make vehicle-based phones, including those on the train, obsolete. With its "handoff" feature, the Metroliner system has been touted by Bell System diehards as the first truly cellular system.

Yet the true cellular future was halfway across the country, in the hands of law enforcement, not aboard a passenger train. The portable radio system being used by the Chicago Police Department—conceived, designed, and built by Motorola—was much closer to modern cellular telephony than a forty-pound automobile phone attached to a railway car.

FROM LOSER TO LEADER

Quartz Crystals

Not everything I worked on was as successful as the IMTS project. In the same period that I was enjoying that success, I was embroiled in an enormously expensive and frustrating quartz problem.

In the spring of 1961, Bill Weisz asked me to investigate what he sensed was a growing problem in the Communications Division's two-way radio business. Bill, who had become vice president of communications products, was proud of the rapport he had with key customers. They knew how intense he was about Motorola's reputation for technical superiority and they also knew that he was an objective listener.

Bill had lately heard customer complaints about deteriorating performance in their two-way radio systems. Radio conversations were noisier than usual, and technicians were being called upon too often to fix the radios. Our field force reported that they believed the crystal oscillators in our two-way radios were not working right.

When you want to listen to a broadcast radio station on your home or car radio, you tune it to that station's radio frequency. In the 1960s, you did this by turning a knob that moved a dial indicator to the number representing the station's frequency. WGN in Chicago, for example, uses the frequency 720 kilohertz. If you were slightly off frequency, the sound would reduce in volume and, as you were further off, it would get noisier until you lost it completely. Modern consumer broadcast radios search for radio stations digitally and precisely lock on to a desired station.

Two-way radios need to be tuned much more precisely than a broadcast radio. The frequencies used by two-way radio stations are crowded together so closely that using a knob would be impractical. A tiny movement of the knob would pass through many stations. Tuning is thus accomplished using a device called a quartz crystal oscillator that, at least theoretically, electronically locks exactly on to the frequency of the desired station. The quartz crystal (also called a resonator) is the heart of the quartz crystal oscillator; it's manufactured to vibrate at a single exact frequency. If for some reason the quartz crystal deviates from its natural frequency, the signal is lost, and the customer loses communication.

Most of our quartz crystals were made in an assembly line in the Augusta Boulevard plant in which my engineering department was located. Bill asked me to evaluate the performance of our crystal manufacturing facility. Was this a case of insufficient quality control, he asked, or was the design of the crystals defective?

The quality test for the stability of our quartz crystals was very simple. We took a sample group of crystals from the production line and heated that sample far above normal operating temperatures for a month. This stressed the crystal to the equivalent of years of use at normal temperatures. We then measured the crystal's frequency and observed whether it had changed over the period of the test. A perfect crystal would not change at all.

I examined the data and concluded that there was no problem. Our crystals were performing in their tests the same as they had been over the previous years. But I turned out to be dead wrong. We had been in trouble for several years and just didn't notice that the quality of our product was slowly deteriorating.

There are few substances on Earth that are more plentiful than quartz. Over the years, it has acquired both technological ubiquity and, somehow, spiritual mystique. I can't speak to the mystical effects that some people ascribe to quartz, but, technologically, it is a fascinating material.

Quartz is piezoelectric, ferroelectric, tribo-luminescent, highly elastic, and often, in its natural form, beautiful. The two characteristics that make quartz technologically valuable are elasticity and piezoelectricity.

If you bend a thin piece of quartz without breaking it, you will observe two interesting things. First, it returns to its original shape and does so very efficiently; there is hardly any heat generated, in contrast with most other substances. We say, then, that quartz is almost perfectly elastic. Second, the act of bending the quartz bar can create a voltage across its faces. Further, if one applies a voltage to the bar, the bar bends. That property is called piezoelectricity.

These two characteristics make it possible to fabricate pieces of quartz into what are effectively small tuning forks, which can be made to vibrate continuously at a very precise frequency. These quartz crystal oscillators (which are about the size of a large grain of rice) are the heartbeat of most computers, cell phones, and electronic watches.

In a two-way radio, quartz crystal oscillators in a transmitter and a remote receiver that "speak" to each other operate at exactly the same frequency so that the receiver detects only that transmitter's signal and not others. If either oscillator deviates from the frequency of the other, the received signal can weaken and, with enough deviation, disappear. A customer would then experience poor performance or a nonworking radio.

What was our problem? The quartz crystals were sealed in tiny metal enclosures to protect them from the environment. As it turned out, the seals on our crystals were not uniformly reliable—oxygen and water vapor were leaking in, damaging the crystals and electrodes. In time, this modified the frequency of the crystal, disrupting its communicative properties, thus leading to complaints of noise from customers. Our initial tests had looked only at the crystals themselves, not the enclosures, so they didn't detect the problem.

It only took a tiny amount of deviation to hurt the performance of our two-way radios, and customer complaints had been incrementally increasing over the years. Yet until the complaints filtered up to Bill, and I looked at the whole picture, nobody in Motorola had put things together and identified the problem. And the problem was serious: disruptions in two-way radio performance were causing our customers no end of trouble. We were spending huge amounts of money responding to complaints, adjusting frequencies, and replacing crystals. Even worse, our competitors were doing better than we were. If we continued down this road, Motorola would be facing financial ruin.

I took immediate steps to stop the bleeding. We started to add some helium to the pure nitrogen that the crystal was immersed in. We then used a device called a mass spectrometer to sniff at the outside of the enclosure to determine whether it was leaking. An astoundingly large percentage of our enclosures were, in fact, leaking, and another sizable percentage had been contaminated before the enclosures were sealed.

Our collective engineering myopia, focusing only on the technical design issues, was contrary to every principle of Motorola's management. Our philosophy was always to approach a product or service from the customer's point of view. It's not enough to test a narrow technical aspect of something and conclude that everything is fine. Looking at something from the customer's perspective means *listening* to your customers and widening your perspective to include theirs.

At Motorola, understanding the customer experience and anticipating customer problems before they happened was foundational. It was in the first weeks after I moved from research to product development that I really started to absorb this core tenet. Since two-way radio was Motorola's bread and butter product, every member of the product team leadership had a two-way radio installed in his car. I installed my equipment personally. The installers on our customers' staff were the people who first opened the boxes in which our products were shipped, and if the installation wasn't perfect, the end user would know it and complain. Every day, as I drove to and from work, we would talk with each other and with a mobile operator, listening for defects in voice quality, for interference caused by poor design or eccentricities in a specific car model. We would try out new models and experimental features. Testers included *every* manager, including the division head.

Engineering field trips involving engineers and management happened daily when we were exploring experimental features or responding to a customer complaint. Our objective was to get into the customer's head, to encounter the unique aspects of the customer's job and anticipate his or her needs. The same thing was true of the production process. I knew every foreman on the lines that built my products and many of the people who worked the production line.

To solve the quartz problem, I had to return to this fundamental tenet.

The long-term fix was a complete transformation of the quartz crystal manufacturing process. We built a "clean room" that was completely sealed off from the rest of the factory. The clean room was pressurized with air that was cleaned with what are called absolute filters so that no moist or contaminated air could enter, and humidity was reduced to minimum levels. Most importantly, key processes involved in building the crystals were isolated even from the people in the clean room. Instead of sealing the crystal enclosures with solder seals, we created enclosures that could be welded.

This was extraordinarily costly, but, in the end, we saved huge amounts of money in customer satisfaction and elimination of warranty repairs. Over a period of eighteen months, Motorola crystals were transformed from being the worst in the industry to the best. The engineers and production line employees responsible for quality and delivery had been demoralized by customer complaints and the manufacturing overhaul. Now, with an industry-leading clean room and vast improvement in quality, they were proud and energized.

When an organization faces an existential problem like ours with quartz, the initial temptation is to manage the external fallout ("crisis communications") and put Band-Aids over the problem. It might seem cheaper to do the short-term fixes and leave the (more expensive) long-term fixes for another day. But this only digs the hole deeper. That's where things were when Bill Weisz asked me to have a closer look. He knew we had to address the deeper causes as soon as possible.

Motorola was stronger for it.

After we revamped the manufacturing process, we set about ensuring that we had a reliable source of high-quality raw material to manufacture quartz crystal resonators. These devices are made from pure quartz. It's possible to find suitable quartz in natural deposits, but pure pieces were hard to find, and the labor needed

A cluster of natural quartz.

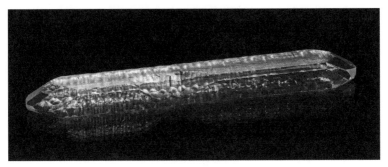

Cultured quartz.

to mine the crystal was becoming costly. The solution was to artificially mimic the process under which natural quartz was produced. This "cultured" quartz was produced in an autoclave, which resembled a large gun barrel. Slivers of natural quartz that looked like long toothpicks were hung in racks in the autoclave.

Small pieces of natural quartz were loaded into the bottom of the autoclave, which was then filled with water, sealed, pressurized, and raised in temperature to reproduce the conditions that generated natural quartz.[1] Molecules of quartz would then deposit on the seed, which, over a period of two to three months, would grow to many times the original size.

Natural quartz is estimated to have grown at about one layer of molecules per year; it would take a million years to make the bar of quartz shown on the previous page. We found an entrepreneur in Carlisle, Pennsylvania, who was manufacturing small amounts of cultured quartz. We bought the company, hired the entrepreneur, and increased capacity by several times.

As a result of these investments—which were painful to make—Motorola became one of the leading global suppliers of quartz crystal material.

FROM MOBILITY TO PORTABILITY

I n 1960, the Chicago Police Department was in crisis. In January of that year, the Summerdale scandal had come to light. The "Babbling Burglar" had told all—eight city policemen had conspired to run a burglary ring. The level of corruption this scandal exposed "was too much for even Chicago's tolerance of misconduct," the *Chicago Tribune* later wrote.[1] Looking for a "new broom and a clean sweep," Mayor Richard J. Daley brought in a new police commissioner.[2]

Daley's choice was unusual but inspired: O. W. Wilson was the dean of the School of Criminology at the University of California, Berkeley. Wilson had initially supervised Daley's search for a new commissioner, and then he ended up as the mayor's selection. Wilson was a "lean, hard-bellied, soft-talking" policeman who'd started his career as a patrolman before skyrocketing to the rank of police chief by the age of twenty-five. Before he left for academia, he'd been a reformer, having built a reputation by cleaning up a bootlegging mess in Wichita, Kansas, after two police chiefs had already been ousted during Prohibition.[3]

In Chicago, as superintendent (the title was changed from commissioner), Wilson quickly made a slew of improvements. A top priority was upgrading equipment for police officers. In the United States' second-largest city, Wilson wanted his patrol officers to police the city's hundreds of beats from police cars where they could reach any location in their beat while remaining in constant contact with

*Police officer with a Motorola HT-220 Handie-Talkie
two-way portable radio, 1969.*

the new and much-touted communications center that opened at police headquarters early during his tenure.

By 1965, five years into his tenure, Wilson had changed his view. Patrolling from squad cars had helped the police department cover more ground and stay in touch with headquarters. But it also isolated patrol officers from the densely populated communities they served.[4] Wilson wanted to get his officers back on their feet walking the street—but still in constant contact with the communications center. To figure this out, he turned to Motorola.

"We've put the policemen out there in cars and now they're losing touch with the neighborhood, with the people," he told us. "Isn't there some way we can get the policemen on the street without forcing them to go back to their cars to talk to dispatchers and to get help?"

Wilson needed something that would work both as a portable handheld two-way radio when patrol officers were on the beat and equally well as a two-way radio when they operated in a police car. At that time, such a thing didn't exist.

Our confidence in taking on the CPD's challenge relied in large part on our extensive experience with pagers.

The invention of the pager is credited to Al Gross in 1922, but he apparently never did much with his invention. In the early 1950s, inventor Richard Florac introduced a primitive paging system. Every subscriber was assigned a three-digit number and carried a receiver to which they would listen from time to time. An operator would orally repeat a list of numbers. If a subscriber heard her number, she would find a pay phone and call the operator to find out who was calling.

In the mid-1950s, led by Bill Weisz, Motorola developed a low-frequency pager—what many called a beeper, which received its signal not from a radio broadcast, but inductively from a long loop of wire that was strung around the building where the pager was used.[5] The pager had to be close to that wire to receive its signal. The advantage of this approach was that it was not necessary to get a license for the radio frequency since the signal from the wire loop traveled such a short distance. One of our first customers was Mount Sinai Hospital in New York City. We outfitted doctors and nurses with pagers so they could be reachable as they moved around the hospital.

Minutes can be crucial when a doctor is urgently needed by a failing patient. But requiring that a doctor remain close to at-risk patients was costly and burdensome for both doctors and the hospital. The pager increased freedom of movement and saved lives as a result.

The pager did nothing more than beep when an operator wanted to get the attention of its owner. A doctor or nurse would then call the operator using the hospital's telephone system, and the operator would pass on a verbal message. The pager would work only if the user and the pager were inside the hospital and reasonably close to the inductive loop.

The concept was simple; the execution was abysmal. The pager used both transistors and subminiature vacuum tubes. Transistors were still in early development stages and performed poorly. Traditional vacuum tubes consumed too much electricity to work by battery power and were unreliable even in fixed applications like radio and TV. In a portable device, like a radio pager that had to withstand drops to a floor and worse, tubes were barely up to the job. The engineering team adapted newly designed subminiature vacuum tubes that used lower power than traditional tubes to serve as the electronic heart of the pager. The engineers had no experience with these tubes but were desperate to lead the market for what they were certain was a powerful communications tool.

The size and weight of the low-frequency pager were at least four times that of a modern cell phone, but it was not robust enough to survive rugged human use within a hospital. When it worked, however, it proved to be a godsend to the medical staff. And the two hundred pagers that were deployed at Mount Sinai worked—most of the time. The pager, as predicted by Weisz and his team, rapidly became integral to hospital operations.

Then came the failures. The subminiature vacuum tubes couldn't survive drops; the loop on the roof didn't provide a signal everywhere in the hospital, and a second loop inside the hospital didn't completely solve that problem. The engineers and the local sales team did their best to resolve the complaints of doctors who missed pages. Even though their old loudspeaker system was far less reliable, the doctors wanted perfection and Motorola didn't deliver.

The final straw was when a doctor, in his frustration, threw his pager against the wall.

Weisz didn't hesitate. Motorola apologized to the hospital and proposed to remove the entire system and refund everything the hospital had paid. We would re-engineer the system and, when it was sufficiently reliable, come back and re-install it.

The hospital rejected the offer. Nurses and doctors refused to give up their devices. The paging system had become so important to the operation of the hospital that they simply could not give it up despite its unreliability. They wanted the system fixed but would use the current system until we delivered its replacement.

The Motorola Pageboy pager, introduced in 1964.

Once people experience the freedom and value that connection and portability deliver, they can't go back. They refuse to go back. People aren't just mobile—they want to be connected while being mobile. That requires portable devices. The idea of mobility *and* portability was becoming a fundamental part of Motorola's culture.

In 1965, Motorola introduced the Pageboy, which worked over radio frequencies rather than the inductive loop used at Mount Sinai. It was mostly used by companies to reach out to their employees. Pagers were "beeped" by an operator, who could "beep" one or as many as hundreds of people in a local area.

Our experience with pagers and our continuing work with portable two-way radios primed us to address the Chicago Police Department's communication needs.

My engineering team, headed by Bob Walker, took on the Wilson challenge. They were already working on a new handheld transmitter-receiver, the HT-220. It was smaller and better performing, with longer battery life, than its predecessor, the HT-100. The HT-220 transceiver was, like prior models, a complete two-way radio with microphone, speaker, and antenna included. When you were talking to someone by radio, you held the transceiver close to your mouth; the speaker was loud enough so you could hear a voice with the radio on your belt. The built-in antenna projected from the top of the unit.

The commercial version of the HT-220 wouldn't be widely available until 1969. To meet the needs of Wilson and the Chicago Police Department, we created a special version. The microphone, speaker, and antenna in the CPD version were removed from the transceiver and encased in a plastic housing the size of a typical microphone and equipped with a specially designed clip that allowed this unit to be attached to a patrol officer's shoulder.

Patrol officers would wear the radio equipment whether in the car or on the street. Their hands were always free when listening, but they were required to press a transmit button to talk. When in the car, the antenna on the officer's shoulder was in a position at the level of the car windows, which were transparent to radio frequency energy (and, of course, to light). If we had left the antenna on the radio, 95 percent of its transmitting energy would have been absorbed by the car doors.

This was deceptively simple, but only part of the solution. The land-based transmitters and receivers of police department systems were designed for two-way radios mounted in police vehicles with high power transmitters and efficient radio antennae mounted on the vehicle. The portable radio on the patrol officer's belt used a rechargeable battery that could only provide energy for relatively low power.

The HT-220, which represented a leap in Motorola's portable two-way radio technology, paving the way for handheld, portable phones.

The need to preserve battery power motivated us to keep the handheld power output low. Our system engineers created an equally simple solution to the power problem. They divided the city into twenty districts, coincident with the CPD organizational distribution, and assigned different frequencies to adjacent districts.

Police departments had long been at the frontier of adopting new land mobile technology, and Motorola had helped them innovate. It was for city and state police forces that Motorola had developed the first two-way portable radio systems. Now, with Chief Wilson and the CPD, Motorola had also created a cellular system. We hadn't conceived the concept of cellular—that originated inside Bell Labs in the late 1940s in an internal memo by D. H. Ring. In that document, Ring never actually used the term *cellular*, but he described some of the basic ideas, such as dividing a larger area into smaller ones (what became known as cells) so that radio frequencies could be reused within the larger area.

The system that Motorola created for the CPD had all of the elements that Ring described. The only thing it lacked from what today we call cellular was handoff, the ability to maintain a call when moving between districts (cells). Ring's memo implied, but did not discuss, the notion of handoff.

With the CPD system, handoff wasn't necessary. Police communications are highly disciplined and efficient. The average length of a complete conversation is about fifteen seconds. It was unlikely that a patrol officer would be moving from one district to another in so short a time.

In responding directly to a customer's need, Motorola had created the first workable cellular system in the United States.

In the short term, however, none of our innovations guaranteed that we would win the actual contract to build the system for the Chicago police. Sure, we were sitting at the top of the hill with an international reputation for quality and breadth in our two-way radio products. In Chicago, we were the hometown team, locked in to the CPD. We had supplied Wilson and the CPD with the new dispatch and radio system in 1961, the largest order the Communications Division had ever received.[6] Our salespeople called upon every level of the department and dined with them regularly.

Yet we had plenty of competitors, working hard to pull us down. When you're at the top of the hill, you're the preferred target, and you can't let your guard down for a moment. We were particularly worried at that time about General Electric coming out with a similar product to our proposed portable police radio. We knew that Wilson had made his request not only to us but also to our competitors, including GE. Complacency was not an option.

That summer, Wilson was going to address the annual Associated Public-Safety Communications Officials (APCO) conference, which was being held for the first time in Chicago. The contract for the portable communications system had not yet been awarded. We didn't want our competition to be aware of our solution; we were keeping it a secret until the award occurred. On the other hand, the possibility that a competitor, particularly GE, would announce a solution like ours before we did was unacceptable. A conference in our hometown was the worst possible place to get upstaged.

As we had done on earlier occasions with other product releases, we wrote and printed catalog sheets for each of our new products, complete with detailed specifications, even though these radios and systems were still being designed. Our public relations people prepared press releases with blank dates. We were ready. If there was a whisper of a GE or RCA (Radio Corporation of America) product announcement, we would pull the trigger and outmaneuver them.

One morning, shortly before the award of the portable radio contract, I was pulled over for speeding by a Chicago cop while on my way to work. In desperation, I tried the friendly approach.

"I work for Motorola," I told the officer. "We make the two-way radios in your cars."

"Yeah!" said the man in blue.

Well, at least I had his attention. "We're designing a new product just for the CPD," I continued enthusiastically. "You'll be able to carry your radio on your belt and never be out of touch."

His reaction was not what I hoped.

"Do you realize how many things I have on my belt now? I got the billy club, I got handcuffs, I got the ticket book, I got the flashlight, I got a gun and spare ammo. Just what I need," he added as he handed me my speeding ticket, "another gadget." I got the ticket, and another worry. Would the police officers who were our users reject the portable radio as a tool of their art? Our job was to get inside the mind of our customers and understand their problems better than they did. We were pretty sure we were making something that would make the officers' lives safer and more productive. The priority, though, was communication and portability. After all, our customer was the CPD and Wilson.

As it turned out, as much as our competitors wanted to win the contract, none of them announced anything new leading up to the APCO conference, and Motorola won the CPD order. We would be a partner in Wilson's quest to keep the police closely connected to headquarters and its communications center, also supplied by Motorola, as well as the communities they served.

As for the belt problem, it soon fixed itself. Shortly after officers were equipped with the new radios, a police officer interrupted a burglary and broke his leg while chasing the perpetrators. He called for backup on his portable radio. Within minutes, the burglars were apprehended and the officer delivered to a hospital. Word got around, and no one complained about the belt problem.

Our contribution helped the CPD fulfill its new motto, introduced by Wilson: "We Serve and Protect."

And we had created a cellular handheld radiotelephone system.[7]

⸻

As a company, Motorola was steeped in this kind of customer and user interaction. We worked relentlessly to understand a customer's business and needs and how we could best serve them. In the 1930s, Motorola (then Galvin Manufacturing) competed with GE and RCA in the two-way radio business. Those companies sold their radios through networks of consumer electronics dealers. Motorola, on the other hand, sold directly to customers and trained its salespeople in technical knowledge to understand the customers' businesses.[8] The

A World War II-era advertisement, by Galvin Manufacturing Corporation of the "Motorola Radio handie-talkie," the SCR-536.

Motorola sales force spent a good deal of time trying to understand the communication needs of, for example, police and fire departments.

In 1940, Paul Galvin, the founder of Motorola, sent an engineer to observe war exercises by the US Army. The engineer noticed how much soldiers struggled with their cumbersome communications equipment. War was on the horizon, and Galvin formed a new engineering team to figure out a solution. What the company developed was the SCR-300 backpack two-way radio, the "walkie-talkie." Later, a handheld version was produced, known as the "handie-talkie"—this was the HT series that formed the basis of the CPD system. The SCR-300 was used widely during World War II in both Europe and the Pacific, credited by American generals as essential to coordination and victory.[9] Later models, especially the SCR-536, were in widespread use during the Korean War.

Thanks to product development experiences like this, Motorola's culture came to revolve around key questions: How can a communications device best serve the needs of a user who is out and about, moving around, and who needs to be in constant touch? In other words, how could portability be added to mobility? People want to move around; they *need* to move around. This means their communications devices need to be portable, to move around with them. This was the charge and the legacy entrusted to me in 1965, when O. W. Wilson approached Motorola with his patrol officer challenge.

During a trip to Phoenix, where several of us were meeting with Motorola's Semiconductor Division, Bob Peth pulled me aside. "We are thinking of promoting you to be product manager for a group we're forming to concentrate on portable products," he said. "I'd like to get your reaction."

I tried to look cool—not my strength under the best of circumstances; I've never been a good poker player. I couldn't hide my excitement. My heart beating like a pile driver, I managed to say, "What's a product manager?"

"We think that the right person to fill that role would be someone who understands engineering but also knows what customers want, can independently run a

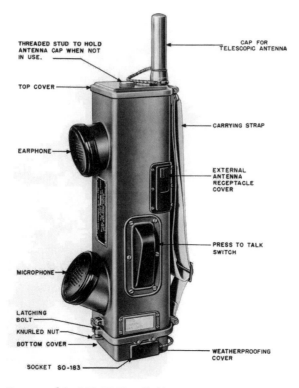

Diagram of the SCR-536 handheld two-way radio, circa 1940.

business within the division," Bob explained. "That person would have profit and loss responsibility and would control product marketing as well as engineering. John and Bill think that you've proved yourself enough to run a new product group called portable products." To have gained the confidence and trust of people like John Mitchell and Bill Weisz was a thrill.

Then Bob said, "I'm not so sure."

For a moment, my thrill deflated, I understood his concern. Bob was a reserved and serious engineer. He was suspicious of any new technology, and he was often right. In our work on the IMTS radiotelephone, he had been correct to stick with vacuum tubes in the power amplifier. He was suspicious of my perpetual enthusiasm.

"What do you think you'll do if we give you the job?" he continued.

"Well, I don't know much about our portable product line, but I know I can hire and motivate the best team, and I'm sure I can learn about the products and the technology." I could tell that he was totally dissatisfied. He wanted me to talk about specific things that I could do in the role he was describing, and all I could think about was how wonderful it would be to run my own operation, with my own team, and with the power to pick my own products. I was steeped in the Motorola culture around portability and wanted to keep pushing our technology forward. My naïveté was boundless.

Despite Bob's reservations, I got the job. I was thrilled, not only by the new challenge but also by the opportunity to be able to respond to customer needs and marketplace demands with creative technology. As the first product manager of this new group, I worked with other members of management to handpick our engineering and marketing teams. The general understanding was that portable products would ultimately be the future of the Communications Division.

The technology part of the job was incredibly enjoyable. We had a spectacular engineering team including chief engineer Bob Walker and Richard Carsello and Norm Alexander, who managed pagers and portable two-way radios, respectively. I roamed the laboratories seeking insight into how and why engineering decisions were made and how portable products differed from the mobile products I had been developing. I met regularly with the marketing team headed by Tom Kain and including Dave Michalak and others who taught me about our customers, how we made our pricing decisions, and how we dealt with Motorola Communications and Electronics, our sales force.

I tried to embody the spirit of Motorola, getting inside the customer experience and understanding the needs of users as much as possible. One way to do that was to experience our product the way it was used. That's how, courtesy of Harley-Davidson, I found myself one day listening to my wife's aunt scream into my ear.

In the mid-1960s, Harley made nearly all police motorcycles in the United States—and Motorola made *all* the two-way radios that were mounted on these bikes. Every year, Harley shipped us one of the new models so we could determine the cabling and mounting for the radio. The bike was always gorgeous, loaded with chrome levers, knobs, and foot pedals. My dream of cruising down highways with the wind in my face finally came true.

I spent my second weekend as product manager for portable products reading the entire instruction manual for the motorcycle. Then I took our bike for a spin around the empty Motorola parking lot until I mastered the gear shifting by the right foot, acceleration by the right hand, and braking by both hands. I then drove the motorcycle home on Chicago's streets and highways. Considering my natural lack of coordination, I was risking my life and that of surrounding motorists, but I was in heaven.

A few hours later, I had convinced Aunt Ann, who was in her sixties and lived a few blocks from my Skokie home, to take a ride with me. At first, she was more excited than terrified as we took a short drive around the neighborhood. That is, until we needed to turn and leaned in to round a street corner. I realized I couldn't hear the Harley's roaring engine anymore. The bike was fine, but screams from my passenger drowned out all other sounds. She was sure we were moments away from being crushed under seven hundred pounds of steel machinery, and we probably were. That was the only time I remember convincing anyone to take a ride before Harley asked us to return the bike a few days later. No doubt, my wife Barbara had no interest in risking her life; she already had more than enough on her plate with our two young children, six and eight years old at the time.

Less death-defying, but no less important for product development, was our team's work in trying to make handheld radios sturdy enough. People aren't just mobile—they're rough on their communications equipment. Experience was teaching us that adults like police officers were harder on radio equipment than children are on toys. Few people expected the Chicago cops to be swinging their radios around by the cords, ripping them out of the receiver. But they did. When a radio subsequently broke, it was our fault, not the officer's.

Norm Alexander took on the challenge of making them sturdier. One morning, he dropped into my office with a proud smile on his face. "Marty, we've finally fixed it. You'll never be able to pull the cord out of this new microphone we've built!"

He'd barely put it in my hands when I ripped the cord right out.

"If we're going to solve this problem, Norm," I said, handing him back the broken equipment, "we really have to *solve* it." A grumbling Alexander went back to the lab and designed stress reliefs in the radio and shoulder-mounted unit that were strong enough to let him do chin-ups supported only by the equipment.

C*⁀⊃

If you truly hope to experience a product or service from the perspective of an end user, you need to get into the mind of the user. That may sound obvious, but it isn't the way many companies operate.

At Motorola, we were working toward a future shaped by portability and mobility. Our vision wasn't perfect; we also continued to make traditional vehicle-based equipment. But we were being pushed, by our customers and our experience, in a different direction. AT&T, however, was stuck in a culture that disdained customer input—its people were "derisive" toward customer desires. This was clear in Bell's dealings with business customers, who went to AT&T with requests about system connections and configurations:

> AT&T would say fine, we've studied your needs, this is your system. And they'd reply: Well, that's fine, but we'd like to see some alternatives, or we'd like to see something changed here or changed there. And AT&T would say: No, we've studied your needs, this is your system. That was the attitude; they did not respect their customers.[10]

Motorola's approach was grounded in getting inside its customers' problems and understanding things from their point of view. Our experiences in developing, testing, and improving different products had imbued us with a technological perspective grounded firmly in observations about human behavior. The most important was that people are inherently, fundamentally mobile. That became my mantra, and I still ardently believe it.

Our work with O. W. Wilson and the Chicago Police Department was one example of this. Meanwhile, our team continued to work on advancing pager technology. Even as we had introduced the first Pageboy, in 1965, our team had never built a high-capacity paging terminal like the one needed to make the new paging system work. The radio pagers we were manufacturing were used in small systems covering a building or a relatively small area and serving a single entity such as an office or a hospital. In those, it was only necessary to select one of ten or perhaps up to a few hundred pagers.

Once again, our path crossed with the Bell System, which was also exploring the opportunities afforded by pagers. They invited a group from Motorola to meet with a Bell Labs engineering team in the early 1960s. We were informed that the Bell System proposed to offer paging service nationwide. Their customers would subscribe for service in a city but be able to get service as they traveled to other cities. This would require complex switching terminals in each city and pagers that could have one of many thousands of addresses. AT&T would develop the terminals but was trying to decide whether to manufacture the pagers in-house or buy them from others—like Motorola.

This was a reprise of the IMTS experience: we were once again trying to convince AT&T to contract with Motorola for development and manufacture of equipment, this time radio pagers. And we were once again intent on getting the business. Part of our challenge, however, was that we were selling to an entirely new group of players at the Labs; all of our IMTS friends were on other projects or had moved on. The new group included Bob Mattingly, a respected senior engineer who would later lead the cellular radio development department. I quickly learned to listen carefully when he spoke, especially when he started with, "I'm just a dumb farmer from Iowa, but . . ." Something incisive was coming.

Presenting to the group of Bell Labs managers and engineers, I tried to convey that Motorola had more experience building radio pagers than any other company. "We are very comfortable with paging technology," I asserted. Mattingly interrupted.

"I don't know much about paging," he began, "but I do know that no one is currently building a high-capacity paging switch or pagers that meet our standards, and we're inclined to develop these products in-house."

Properly chastised, I humbly admitted that we had little experience with the switching terminals that would be used to send a paging signal to alert one of thousands of radio pagers in a city. But we had been developing and manufacturing pagers for almost ten years and were prepared to share our vision for the future of paging with the Labs.

The prototype pager that we showed them was lovely. It was the size of two ballpoint pens side by side and emblazoned with the Bell logo and the name "Bellboy." (In 1962, Bell had displayed a prototype, the Bellboy pager, at the World's Fair in Seattle.) The prototype could be plugged directly into an AC outlet to recharge its battery. This idea was totally impractical. Adding the plug and charging circuitry would undesirably increase the size and weight of the pager and adding safety features for the electrical power from a wall socket added still more size and weight. But the dummy unit was nonfunctional and looked beautiful. It achieved what we intended, which was to get the attention of our potential customer. Unlike the IMTS experience, the group was only mildly impressed but indicated that they would be willing to continue the conversation.

Back at Motorola, John Mitchell put out the word: we were going to continue cooperating with Bell Labs. Regardless of the Bell decision, we would stay the course and build the paging product line.

AT&T thought paging would be a substantial business: their design called for a capacity of three thousand pagers, enough to handle a large city. They later increased the capacity of their proposed terminal to six thousand pagers. While the Bell System had a family of telephone switches from which they could draw to modify into a paging terminal, we had no such basis for our product. Instead, our team designed the terminal around the PDP series general purpose computer that had recently been introduced into the market by the Digital Equipment Corporation (DEC).

John and I were much more optimistic about the paging market than the Bell people. Perhaps it was due to our experience with Mount Sinai or our work with

Chicago patrol officers. Whatever the reason, we approached pagers from the standpoint of mobility and portability. (We even toyed, briefly, with the idea of leaving Motorola and starting companies to offer citywide paging service.)

New manufacturing technology was necessary to make this happen. We were using printed circuit boards and traditional components—resistors, coils, and capacitors—in our other products, but they were totally unsuitable for a product as small as the pager we envisioned. We investigated a new manufacturing technology in which miniature components were mounted on ceramic substrates that could be manufactured with much smaller circuit patterns than was possible with the printed circuit.

Since we had no experience with thin-film substrates, I decided to find an expert. A search of the literature identified Mort Topfer as just such an expert. I interviewed and hired him. Mort was a superb engineer and turned out to be an equally superb executive, eventually becoming president of Motorola and, later, vice chairman of Dell. For myself, it was becoming clear to me (and Motorola) that I was not traditional executive material. I wasn't interested in managing the details. I wanted to be with the troops inventing things. Although the corporate ladder wasn't for me, Motorola created positions that suited my temperament and inclinations and enabled me to be successful.

The result of the work led by Topfer was the Motorola Pageboy II, introduced in 1971. This would be one of the company's most successful products. At about the same time, we released Motorola's flagship Metro 100 switching terminal with a capacity of one hundred thousand paging customers. This was one hundred times greater than that offered by any other company.[11]

Now Motorola had a different pager challenge on its hands. Largely because of our product development efforts, the Federal Communications Commission allocated several slivers of bandwidth for shared use by paging carriers. We had all the elements of the new business—assuming, of course, that there were customers interested in this new service. But who would buy this monster one hundred thousand pager-capacity terminal when the only service providers were Radio Common Carriers, which were typically small businesses serving hundreds, not thousands, of customers?

Thankfully, the Motorola sales force came through. Don Brickley, one of our sales vice presidents, approached several RCCs in Los Angeles and proposed that Motorola offer shared time on the Metro 100 terminal. Homer Harris was the RCC owner with the vision and risk tolerance to see the possibilities. He led a group of seven other RCCs to accept the offer. Subsequently, Harris's company bought the terminal and managed, with Motorola's help, to increase its capacity even further, to over 130,000 customers. He eventually sold his company to Metromedia, which was in turn purchased by one of the Bell Operating Companies.

Harris and other RCC owners—including Clayton Niles in Arizona and Texas and Larry Garvey in Louisiana—helped build the pager market into an explosively fast-growing one. A new generation of entrepreneurs, such as Roger Lindquist and George Perrin, entered the market and started new companies. (Many of these entrepreneurs later helped build out cellular telephony.) Motorola's Pageboy II and Metro 100 were the foundation of this new billion-dollar industry and new form of communications. The Pageboy II launched a new industry based on citywide and nationwide paging. It took a technology that had managed a few hundred pagers in one building to a sprawling network that accommodated millions of radio pagers, or beepers, across the world.

The Bell System paging business, based on the Bellboy, never really materialized.[12] Even though Bell had access to similar slices of bandwidth as the RCCs, they were unable to compete effectively. Even as paging prospered for decades, attracting over fifty million subscribers in the United States alone, the Bell System never had more than minuscule market share. The Radio Common Carriers, having battled AT&T (and each other) for years, knew how to run efficient and effective local businesses.

In the face of this, AT&T resorted to tactics it had employed in its earliest days to inhibit the business of competitors. Telephone lines necessary to connect radio stations to the RCCs' central office would mysteriously fail, and repair would take hours or days. In Los Angeles, Bell and another company were found by a court to have interfered with the business of a group of RCCs and were prohibited from operating paging systems for five years.

After a forty-year run, use of pagers declined precipitously in the 2000s as the cost of cell phones dropped and coverage and functionality expanded. Nevertheless, a few million paging customers still exist in the United States today, concentrated in the health-care industry. As recently as 2017, a survey of physicians in hospitals found that 80 percent had been provided with pagers and that, for most of them, pagers exceeded other technologies in terms of patient care–related communications.[13]

The long and sustained success of the paging industry was a crucial ingredient in the introduction of cellular telephony. The RCCs initially feared cell phones in the 1970s and 1980s but recognized the freedom offered by connectivity; their businesses had profited from that understanding with pagers. When the later battle between AT&T and Motorola centered on a political and philosophical struggle between monopoly and competition, the cadre of successful companies and skilled communications executives at the RCCs became important Motorola allies.

⌒※◯

Motorola's experience and expertise were based on radio, which was (and is) essential to cell phones. Partly because of this, Motorola's contribution to development of cellular telephony and the handheld portable phone has often been minimized—some said it was computer technology, not radio technology, that provided the true breakthrough.[14]

Motorola, however, was also working on the application of transistors to more and more technology. We hadn't invented the transistor—that was Bell Labs in 1947. But by the mid-1960s, Motorola had become a leader in its development and application. Motorola's 2N2222 transistor, introduced in 1962, became one of the best-selling and most widely used transistors in history.[15] The company had a good idea of the direction that transistor technology was heading. Motorola produced some of the first televisions and radios with transistors and had shrunk transistor radios down to pocket size. We knew that, at some point, transistors would transform phones, making them smaller, more powerful, and portable.

More broadly, Motorola approached mobile telephony from multiple directions and experiences. There was portability, made clear in our CPD relationship. There was the on-demand need, demonstrated by the Mount Sinai experience. There was a growing realization that the markets for mobility and portability would be large, demonstrated by the rollout and adoption of pagers. And there were shrinking transistors. Underneath all of this was a continuing push within Motorola, based on customer feedback, for better-performing and more functional technology.

Motorola had built, for example, a system of radio communications for workers at airports, from gate attendants to those working on the tarmac.[16] As part of the system, we had manufactured elaborate "holsters" for them to hold their two-way radios. Once, while I was walking with John Mitchell through the terminal at O'Hare Airport in Chicago on our way to visit a customer, John stopped.

"Notice the gate attendant and airline employees rushing through the terminal," he said, pointing to several people walking quickly past us. "They're clutching their handheld Motorola two-way radios in their hands so they are readily available, even though we provided them with holsters they could attach to their belts." He was right: these people were continuously using the radios as working tools.

"If we can make our radios small and light enough that people can carry them, the radios will become extensions of their bodies and they will not part with them," he said.

I still think about John's comments over fifty years later, when I see how many people today clutch their cell phones continuously as they move through their lives. I was recently at the hospital for a visit prior to hip replacement surgery. As I left, EMTs were struggling to load a heavily bandaged patient on a stretcher into an ambulance. The patient was pecking away at his phone through the entire process.

John was right—we do not want to part with our communications devices.

The future of technology is not determined only by technology itself. Regulatory policy, legal rules, corporate strategy decisions, competitive relationships— these and more shape technological development. Cellular telephony had been imagined for decades. But actual development into a communications system depended on far more than technological feasibility.

Motorola approached cellular technology and the future of phones from the perspective of a question: How can we marry telephones with mobility and portability? AT&T and Bell Labs approached it from a broader technical perspective in terms of signal strength, location change, handoff, and other such technical issues.

Inevitably, AT&T saw cellular through the lens of a monopoly. How could this new technology—cellular communications—extend the telephone monopoly? Even if AT&T, through Bell Labs, solved the technical challenges, the system would have remained subservient to the telephone monopoly and thus underdeveloped. This was evident in our discussions with them about a paging system. Monopoly necessarily led to a vision that was shortsighted. And, in the late 1960s and early 1970s, expansion of that monopoly to cellular phones was a very real possibility.

How Motorola and AT&T were cooperating and competing on the technology front is only part of the story. Yes, this was a technical competition, a struggle between two companies with different histories and different philosophies. But it was also a battle over the future of communications shaped by those contrasting histories and philosophies. And the two companies were about to engage in full-scale warfare. The theater? Washington, DC.

But first, we'll head to Florida.

A CUP OF COFFEE GOES FURTHER THAN A DROPPED CEILING

Around 1967, not long after I was elevated to the position of product manager, we started planning to move the entire business of portable products—engineering, marketing, manufacturing—to Plantation, Florida, on the outskirts of Fort Lauderdale. That area was expanding rapidly, and we needed to diversify and tap a new source of engineering and business talent. For me, this was a dream fulfilled. I was brimming with ideas to revolutionize our manufacturing processes, improve product quality, and inspire a reinvigorated organization. This was my chance to design, from scratch, the ideal factory and office building that expressed my vision of the future.

I sent my production manager to Florida to open a temporary facility and start building a new team. We engaged an architect whom I deluged with ideas for the factory and offices. I wanted people to be proud of their workplace. The design ended up looking more like a laboratory than a production line. In contrast with other Motorola factories, which had painted concrete floors and exposed wiring and plumbing in the ceiling, we would have tiled floors and dropped ceilings that disguised all those pipes, air-conditioning, plumbing, and electrical conduits. "How can we ask people to make precision, high-quality products in a pigsty?" I thought.

Homer Marrs, a long-serving Motorola executive, was now my division manager. He was opposed to any unnecessary cost and especially my dropped ceilings. I was determined. I had experienced Homer's powers of persuasion and was prepared to resist them.

"This is your operation and you should run it the way you want to," Homer began. His charm kicked into high gear. "I know how smart you are and that you're going to watch every nickel that's spent on your new factory. You'll figure out that a dropped ceiling will never get you a return on the money you spend."

He continued, "How do you figure out if living in a nice environment will make you more productive? Look at all these companies that give away free food. Are they getting a return on that? I think they are overdoing it, living in a dreamland."

Coming full circle, he concluded with a dose of charm: "It's your business and I know you'll do what's right."

Maybe those companies were wasting money on free food, but by my way of thinking, they were attracting the best minds because of it. Similar debates occur today. Some companies are firmly in the amenity-rich camp: turn employees into team members who feel like their contributions are recognized, and they'll pay you back in productivity. Other companies are at the opposite end: no frills, few perks, but lots of performance-based incentives.

For our new facilities in Florida, I resisted Homer and got the dropped ceilings and tiled floors. I believed the physical environment of employees directly affected profitability. This was the kind of thing that mattered, but in a different way. Most companies (at the time, at least) used their physical space to differentiate lines of authority and position. When I first took over portable products, prior to the Florida move, our group was assigned to a newly renovated area within the Augusta Boulevard headquarters in Chicago. I had a private office that had once belonged to Elmer Wavering, who was now vice chairman of the company. (Wavering had been an early Motorola employee and helped developed the auto radio, the company's first big commercially successful product.)

The fact that the one window in the office looked out onto the garbage shaft was irrelevant; there was genuine carpeting on the floor, albeit a bit tattered. When I was at Teletype, someone's carpet was ripped out of his office because he was one level too low for such a perk. Motorola, in general, had little appetite for such nonsense. I felt special, and I wanted my colleagues to feel special. The first Saturday after my promotion to portable products, I purchased and hung a half dozen hanging pots containing artificial flowers throughout my new engineering laboratory.

It was a small move, but I wanted to communicate the sense that we were all in this together. I wanted my team to like where they worked and enjoy being there—and, hopefully, to be productive as a result.

In Florida, in addition to changes in the ceiling, we experimented with different ways of enhancing collaboration. The new factory had no reserved parking except for visitors; why should the executives have reserved parking when it rarely rained and never snowed? All parking spots were within a few minutes' walk, and everyone had the same walk. The marketing, engineering, and financial people were located a short distance from the production lines. My office was at the center: we were set to work as a closely knit team with all of us immersed in the whole process.

The biggest challenge was trying to replicate and extend Motorola's corporate culture of objectivity, intense involvement in the technology, and the primacy of work in one's life in a physically distant place—and a very different place, where a junior engineer had a pool in his backyard or a surfboard in his car. The challenge was compounded by the fact that the person running the new Florida business (me) was not fully present there. For the first year of the Florida relocation, I remained in Chicago while we hired employees in a temporary facility in Plantation.

Not unreasonably, the new Florida employees adopted a vacation-style culture. There was no one with them to cultivate and foster the Motorola culture. When I arrived full-time in Florida, I thought I was back at Teletype. The Florida facility was afflicted by "swinging hanger" phenomenon. At five o'clock, everyone was gone, the coat hangers left swinging from their swift and punctual departure. I was devastated. Didn't my managers realize how unique our nerdy, enthusiastic engineering culture was, and how important it was to maintain the essence of that culture if we were to retain our competitive engineering leadership?

A dropped ceiling couldn't compensate for the absence of elements that truly foster a culture of collaboration and productivity. I had failed to recall the experience of my early days at Motorola, when leaky roofs were no barrier to innovative thinking. Organizational culture is much more than the physical surroundings.

I took for granted that my appreciation for the way we operated was shared by everyone. The swinging hangers taught me that the behaviors that create a leadership corporate culture have to be continually reinforced. The most valuable reinforcement is from the top. Team dedication to a mission and buy-in to their roles in that mission aren't enough. Example, not just words, is needed to demonstrate what makes an organization different—and better.

Fortunately, I had a mentor who led by example in John Mitchell.

John was an extraordinary and intense teacher. His talents for observation and understanding of the basic needs of people were exceptional—he took his teaching role seriously and was my mentor for most of my years at Motorola.

As an engineering manager, I was intimately aware of the budgeting process, but I knew little about the profit-and-loss (P&L) statements and return-on-investment balance sheets by which financial performance of products were measured.

"Now that you are managing a business and not just an engineering group, you need to understand the importance of the profit-and-loss statement as a management tool," he told me. "The titles on each column or line are superficial. To do your job properly, you need to know how each number is derived, who defines that line, and what their criteria are. You need to know which numbers you control directly and which you can influence indirectly."

He offered me insights into how to use the P&L as a management tool in dealing with my team and with the corporate service groups on which I depended but that did not report directly to me.

We went through a current statement one line at a time. He showed me that the only costs I could control directly were the salaries of my people; even the overhead charges associated with their salaries were out of my control. Most of the charges on my P&L statement were allocated based upon obscure formulas over which I had no visibility or control. But each of these charges represented services that I was ostensibly receiving.

"The challenge," John said, "is to get your fair share of these services. Many of the lines on the P&L statement represent people whose talents and services you need to do your job. If you treat these people respectfully and cultivate your relationship with them, you'll get your fair share, meaning more than those who don't pay attention."

This lesson was burned into my mind and became a crucial element in the development of the cell phone.

"Notice also," John continued, "some costs are labeled as variable and others as fixed. The theory is that as the volume of shipments of a product increases, variable costs—labor and materials, for example—will increase proportionately; in contrast, accounting services, personnel, and other costs remain constant. If the fixed costs were truly fixed, the profit margins would go up dramatically. This rarely happens, because managers use the increased volume as an excuse to hire more staff, which increases overhead, and they spend money on expenses that are often unnecessary. Only by aggressively managing the few expenses that you control can you influence your profitability."

He continued, "The bean counters decide what your allocations are going to be. Although you can't control the allocations, you should be aware of them, and if they're not fair, you can negotiate them. The better your relations are with the people in the financial organization, the higher your chances of getting your fair share."

John's financial lessons were supplemented by wisdom from Bill Weisz about product pricing. Bill began working at Motorola in 1948, would become president in 1970, then chief operating officer, and eventually chief executive officer. Price and cost, Bill taught me, are different—in fact, they're unrelated. When pricing a product, cost was only relevant in determining whether the product would be profitable. Price, on the other hand, is determined by factors like value to the customer, competitive issues, customer perception, and price elasticity.

In addition to the P&L, managing a product development business within Motorola meant managing more people, both directly and indirectly. "Good managers can extract more productivity out of their people without using their authority or badgering them," John told me. He practiced management as a psychological challenge.

John was ruthless in setting goals and standards but sympathetic and charming when he thanked people for their hard work. He was steeped in the Motorola culture of objectivity. Art Reese had taught that to us in the 1950s; Bob Galvin had exemplified it in his consistent behavior. Personality, personal biases, and unsubstantiated opinions had no role in organizational decisions. Religion, race, and skin color were irrelevant; it took a while for the corporate culture to face gender inequities, but when the time came for that, Motorola embraced action aggressively. Business problems were attacked with a minimum of emotion, although the dividing line between emotion in business decisions and passion regarding product integrity was often fuzzy. That doesn't preclude having good relations with people. John Mitchell didn't care at all about being liked, but he did care about being respected.

"If you want people to support you," John told me, "figure out how to make their job easier, figure out how to make them successful while you get what you want. Just being a nice guy, just complimenting them, doesn't work. People need some level of anxiety to be productive. Let me tell you about 'pigeon pecking.'"

John told me about a psychological research group that set up an experiment in which a pigeon would be taught to peck at a paddle, upon which a bit of corn would be released. As expected, when the pigeons had sated their appetite, they stopped pecking. The psychologists then undertook to maximize the number of times the paddle was struck and the amount of corn the pigeons could be trained

My three Motorola mentors (left to right): Bill Weisz, Bob Galvin, and John Mitchell.
At the time of this photo, in 1975, Weisz was president and chief operating officer;
Galvin was chairman and CEO; Mitchell was assistant COO.

to accumulate. If too many pecks were required to get a single kernel of corn, the pigeons became discouraged and gave up. If it was too easy, the appetite of the pigeons was quickly sated, and they likewise gave up. The researchers discovered that they could optimize corn production for every pigeon by creating just the right amount of frustration.

"It's the same way with people," John explained. "We make demands of people and reward them when they fulfill these demands. But if the demands come too often, they burn out and rebel, or they are sated by too many rewards. You can visualize a graph with productivity on the vertical axis and anxiety on the horizontal. People who are completely content are not very productive. As they become more intense about their work, they become more productive, but at some point, they start to fold under the pressure, and at the extreme they have a nervous breakdown and their productivity drops to zero. So, that's my philosophy. Set the goals high, demand the best of your team. When I sense they're getting close to the cliff, I buy them a cup of coffee."

John literally did that. He didn't need to flatter a person. The mere fact that he was taking time out to listen to the details of a person's job and to discuss them was a show of respect. He was magnificent. I tried hard to apply what I was learning. People who worked with me—including those who didn't report directly to me—would later say I pushed them to expand horizons: "if Marty knew it had to be done, it could be done."[1] I didn't push them just for the sake of it, of course. I tried to show them where we were trying to go. As Chuck Lynk later said, echoing Paul Galvin: "You gotta have someone there who tells you to reach out."[2] That's what John was showing me how to do, better than any dropped ceiling ever would.

I didn't get the chance to correct my error with the ceilings. In 1971, not long after relocating to Florida, I was moved back to Chicago. The next town to test my mettle would be Washington, DC.

DAWN OF WIRELESS

The War in Washington that Created Cellular

In May 1973, I tried to calm myself as I stood in the hearing room of the Federal Communications Commission on M Street in downtown Washington, DC. I was mentally girding myself to speak at a hearing on the allocation of radio spectrum for cellular phone service. Much more was at stake than just the technicalities of spectrum allocation. The real question that loomed over the hearing was whether AT&T would be given monopoly control over the proposed cellular radiotelephone service. To me and my Motorola colleagues, this hearing was about the future of personal communications in the United States.

John Mitchell was testifying, as was Karl Nygren, a partner at the prestigious law firm Kirkland & Ellis. Nygren was a brilliant lawyer, hired by Bob Galvin to work on key strategic issues for our company. I would also be testifying to the seven FCC commissioners.[1]

My mission was to counter AT&T's specious arguments promoting their proposed monopoly control of cellular communications. My argument was that AT&T was asking the FCC for far more spectrum than it needed. In doing so, AT&T would prevent others from acquiring spectrum and competing with them in the future. This tactic continues to be a strategy of existing carriers, descendants of the old AT&T, who, despite the fact that they compete, can't seem to shed their monopoly heritage. If the old AT&T accumulated all the spectrum space it was asking for in 1973, the company would have the possibility of using it in any way and at any time, to the detriment of its unregulated competitors. The others from Motorola would speak to our position that a monopoly was unnecessary—that the market for portable cellular communications was big enough to support multiple competitors.

I had rehearsed my comments dozens of times, alone and in the company of Karl and others. When I interviewed him for this book, Karl reminded me that at Motorola's offices, we had set up a near duplicate of the FCC hearing room, where we rehearsed our presentations. Our lawyers would take the role of FCC commissioners, sitting like judges on a raised podium.

My first slide listed bullet points emphasizing a half dozen of AT&T's claims about radio spectrum and cellular service. It was an effective slide, except for the fact that the commissioners could read ahead of my presentation and I'd lose them. So, we decided to paste strips of paper across the bullets, which I could peel off as I addressed them. After the rehearsals, we carefully rubber cemented the strips onto the chart.

In the hearing room, I set up the easel with what now appeared to be a blank pad of paper. This wasn't a demonstration meant to wow the attendees, like the ones we did for Bell Labs—this was a government agency hearing. But we were still trying to sell the FCC on our position. I would need to draw on all the showmanship skills developed with John—and the selling skills I observed in my mother—to make my case effectively.

After several other testimonies, it was my turn. I delivered my memorized introduction and turned to the easel to go through AT&T's transgressions. I ripped off a strip of tape. There was a loud tearing sound; the adhesive that we had applied several days earlier had hardened unexpectedly. I was startled, but so was everyone else in the hearing room, including the commissioners, who had been lulled into near comas by the previous speakers. I had everyone's attention.

"AT&T's request for 75 MHz [megahertz] has hidden reserves built into it." *RRIIIIIPPPP.*

"AT&T is claiming 37.5 percent more spectrum than it needs." *RRIIIIIPPPP.*

"Based on AT&T's own projections, within twenty years it will only be using half of the spectrum it is seeking now—and only within larger markets." *RRIIIIIPPPP.*

"The cost comparisons used by AT&T are only between variations of their own system, not accounting for any competing systems." *RRIIIIIPPPP.*

I proceeded methodically down the list, ripping tape and disproving claims. There should have been no doubt remaining that AT&T was exaggerating the amount of spectrum it needed to provide the services it proposed to offer.

At the end of my testimony, an AT&T executive called out, "Mr. Chairman, I object! This is supposed to be a hearing, not a trial." The chairman, Richard Wiley, responded, "You're right, and that's precisely why you can't object." The ensuing laughter from the several hundred attendees only reinforced my testimony.

When I returned to Chicago, I was called into the office of Bob Galvin, CEO and son of Motorola founder Paul Galvin. My mind was racing—now what had I done? I'd made numerous presentations to Bob over the years but rarely visited his office and even more rarely alone. My apprehensions were reinforced by his stern look as I walked through the door. Bob subscribed to the Clean Desk, Clean Mind theory. His desk was bare whenever he had visitors. On one of the other rare visits I made, I had entered a moment sooner than he expected, and he actually blushed as he quickly shoved a small stack of papers into the desk drawer.

This time, Bob pulled a letter out of his desk drawer and handed it to me. It was from John deButts, AT&T's chairman. It began, "Dear Mr. Galvin, I take great umbrage at the remarks made by your Martin Cooper impugning the integrity of the Bell System." I didn't know how to react. Bob wasn't in the habit of sharing high-level documents with me. Maybe my triumph at the hearing was no such thing.

"How do you feel about antagonizing the chairman of the biggest company in the world?" Bob asked sternly. "And a customer to boot." I struggled to fabricate an answer.

But Bob couldn't suppress his glee for long. He burst out laughing and dismissed me.

Bob's amusement at my hearing performance was a release valve. The truth was that tensions within Motorola around the FCC's deliberations had been increasingly high since the mid-1960s. The commercial future of Motorola hinged on these hearings. During the 1970s, Motorola would go from the peak of business success to the brink of corporate disaster and then to the establishment of a completely new and innovative form of communication. This up-and-down saga involved the FCC, lobbyists, secret meetings, last-second airport dashes, lobster lunches, pin-on buttons, and just a touch of corporate irony.

In the late 1960s and early '70s, Motorola faced an intense strategic dilemma. We were caught between our existing lines of highly profitable business, new possible directions, competitive threats, internal disagreements, and regulatory policy. All of this stemmed from the potential opening up of new parts of the radio frequency spectrum. To fully understand this, and the war we waged with AT&T, let's take a short detour into radio spectrum and its use and allocation.

When you make a call on your cell phone, you have the equivalent of a private radio channel. It's not unlike the physical telephone wires that carry voices on old-fashioned landlines. There are thousands of such radio channels, all of which make up the "radio frequency spectrum"—the spectrum. Different kinds of radio channels within the spectrum are used (or "allocated," in industry parlance) for different types of communication. Police and fire communications, airport radar, the GPS signals that make your map app work—all use the radio spectrum. Closer to home, your baby monitors, garage doors, and Wi-Fi use radio channels. Within the very high frequency (VHF) band of the spectrum, FM radio and CB radio operate. At ultra-high frequency (UHF) and above, you'll find microwave ovens and cell phones.

You might think of the radio spectrum as a roadway system. When you drive your car on a road, you use a segment of a lane in that road. The government may dictate rules that assign certain traffic to specific lanes, but two cars or trucks using the same section of road must be in different lanes or spaced adequately in the same lane or they'll collide. Similarly, a cell phone user is assigned a radio channel, while her cell phone is in use, for an area around her called a "cell." Other users in the cell must be on different radio channels.

Who owns the road, the radio channels? Well, I do. You do. We do. The radio spectrum is public property because it is considered a natural resource. To avoid collisions among users, governments manage the spectrum and allow entities to use parts of the spectrum so long as it is in "the public interest, convenience, and necessity."[2] In the United States, the Department of Commerce manages the spectrum allocated for military and federal government use, and the FCC manages the rest. Within allocated bands of spectrum, the FCC can assign parts to a police department or fire station. Or it can license use of part of the spectrum for radio channels, TV stations, or cellular operators. Theoretically, the FCC can take these channels back if they're not used properly, but that rarely happens. The FCC can also assign spectrum that is shared among multiple users, like Wi-Fi or CB radio.

For one or two decades after Guglielmo Marconi sent the first radio transmissions in the 1890s, radio was mostly an amateur pursuit. The US government laid out rules for the use of radio in 1912, when only a few people owned radios and traffic was low. In the 1920s, helped by radio station broadcasts of election results and boxing matches, radio exploded in popularity. By the end of 1922, there were five hundred radio stations, and a few years later, four million households owned radios.[3]

Traffic was growing; interference and competition for lanes were also growing. First the Federal Radio Commission (created in 1927) and then its successor the FCC (1934) were created to manage the spectrum. The FCC is in charge of allocating and assigning most of the spectrum—determining who can use what roads and what lanes. With a spectrum license from the FCC, you are granted exclusive use of a traffic lane or lanes. Having exclusivity of anything creates scarcity and value. AT&T's aim in the early 1970s was to grab as much spectrum as it could because of that value, and because having exclusive use of large swaths of radio spectrum would have the effect of perpetuating its monopoly.

When Marconi sent the very first radio transmissions, he was basically using all of the available radio spectrum—all of the traffic lanes. As soon as a second person came along to send a radio signal, there was crowding; the radio signals interfered with each other. We would have run out of radio channels a long time ago if engineers had not continuously figured out ways to squeeze more information into each channel. They have been able to divide the spectrum up into separate roads, create more and more lanes within those roads, and figure out how to get the cars in each lane to go faster.

This type of technical innovation has greatly expanded our ability to squeeze more and more traffic onto the radio spectrum. Over the last hundred years, the ability of the airwaves to accommodate wireless traffic has doubled every two and a half years. This steady and relentless growth is described in what I call the Law of Spectrum Capacity, which some call Cooper's Law.[4] It's not really a law, just as Moore's Law is not a law—it's the 1965 prediction by Gordon Moore, cofounder of Intel, that the number of transistors per integrated circuit would double every year as cost fell. These "laws" are observations that a long-standing trend will continue.

In 1993, I observed that, in the fifty years after Marconi's first transmissions, the information-carrying capacity of all radio channels in the world had grown by a factor of one million. Yes, that's right: an improvement one million times greater than the first radio transmissions. And this continued: the next fifty years saw another one million times leap. The millionfold increase in spectrum capacity every fifty years means that, today, "we enjoy over ten trillion times the wireless capacity of networks a century ago."[5]

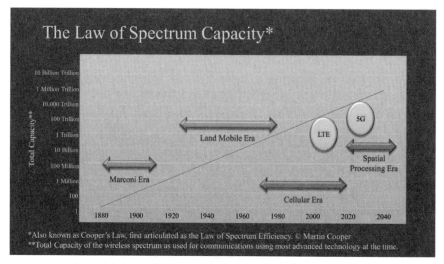

The Law of Spectrum Capacity, also known as Cooper's Law, which is based on my observation of the continued doubling of spectrum capacity every thirty months for over a century.

This is one of the most remarkable engineering feats in human history. It's why I believe we'll never run out of spectrum capacity. We already know what techniques will keep doubling capacity for the next fifty or sixty years. I believe technologists and engineers will stay ahead of demand, just as they always have.

The Law of Spectrum Capacity only works if industry adopts the latest technology. By the mid-1960s, the traffic lanes of the radio spectrum were getting very congested. A 1966 report by the Telecommunication Science Panel, convened by the Department of Commerce, called spectrum utilization a "silent crisis." New ways of increasing spectrum efficiency were "urgently needed." The "situation is critical," the report said, because "it is inevitable that further portions, if not all" of the spectrum "will become saturated."[6]

Growth and usage were especially high in land mobile services—precisely the area of Motorola's largest and most profitable business. In 1965, the FCC chairman had called congestion in land mobile "one of the Commission's most pressing problems."[7] Motorola was feeling this squeeze. The radio channels that customers licensed were the lifeblood of the Communications Division. In fact, Motorola typically helped facilitate customer applications to the FCC for spectrum licenses. The company maintained a small staff in Washington for this specific purpose and worked closely with the FCC. Together with our customers, Motorola formed a lobbying group, the National Association of Business and Educational Radio (NA-BER, pronounced "neighbor"). It wasn't easy to organize our allies into anything like a coherent group; they were as often fighting each other as they were AT&T.[8]

Motorola had an immediate need for more radio channels to serve the two-way radio business. In densely populated areas, there were no unassigned radio channels left in the 150 MHz band—the one in which most of our customers operated. In these cities, all the radio channels in this band were already assigned to our customers and our competitors. Motorola had been lobbying the FCC to open up channels in the 450 MHz band for land mobile. We were asking for two-way radio communications to be expanded from the VHF bands of the spectrum, twenty-five and 150 MHz, to the UHF band, the 450 MHz band. On this, we were loosely aligned with AT&T—at least in terms of who the opposition was.

A decade earlier, in 1958, AT&T had asked the FCC to allot spectrum in the 800 MHz band for the IMTS service—car phones. The FCC had ignored the request, and for good reason. IMTS was simply too inefficient to accommodate large numbers of customers. In 1968, in response to the "silent crisis" in spectrum use, the FCC let it be known that they were inclined to open up more radio spectrum for mobile telephony. The commission published what's called a "docket inquiry," seeking proposals for use of spectrum in the 806 to 960 MHz band.

This attempt to open up spectrum for mobile phone use was enormously controversial. The UHF spectrum bands that the FCC was proposing to reallocate had been allocated in the 1950s for television stations, and they had a powerful political presence in Washington through the National Association of Broadcasters (NAB). Motorola's Washington lawyers told us about reports of local TV stations providing free airtime to politicians in exchange for their support in keeping the FCC from touching their allotted spectrum.

The problem was that television stations never came close to using more than a fraction of it. They had exclusive use of this spectrum, so they never even tried to use more of it. There was simply no incentive to do so; instead, the incentive was to hoard spectrum even as it went unused. So-called guard channels were created that supposedly protected TV stations from interference by stations close in frequency. As receiver technology improved, these guard channels were unnecessary. But broadcasters persuaded the FCC to retain them to minimize future competition. Their claim was that they would use spectrum for cultural and educational programming. What they really ended up doing was jealously protecting empty traffic lanes.

To make things worse, the FCC wasn't in complete agreement about this docket inquiry. One of the commissioners, Robert E. Lee (yes, really), opposed giving spectrum for mobile phones. They were only a status symbol, he said, and were an example of companies "frivolously using spectrum."[9] This idea that radio spectrum is limited and should only accommodate uses that are deemed essential and in high current demand shaped the cellular debate for years and continues to influence policy making today.

Motorola and AT&T were thus aligned in our desire to see more spectrum opened up, and in our belief that television stations were hoarding unused spectrum. Several months after it was published, AT&T filed a response to the FCC docket inquiry. It was remarkable—in the worst way. What AT&T proposed was ambitious and audacious and arrogant; exactly what one would expect from a regulated monopoly with a fixed and limited vision of personal communications.

⌒✳⌒

At a meeting in the middle of 1969, about a dozen members of Motorola's FCC Strategy Committee gathered on the top floor of the headquarters building in an oak-paneled, high-ceilinged, and windowless boardroom. We had just received a copy of AT&T's response to the FCC docket inquiry, which AT&T willingly shared with Motorola because we were, after all, partners on IMTS.

I arrived in a rush, my customary style. I've never figured out why, but no matter how much I plan to arrive early to meetings, I'm most likely to be in a hurry. I had run my morning 10K and driven at breakneck speed to the office with just minutes to spare. I glanced at the day's schedule, grabbed a handful of sunflower seeds, and raced to the top floor. In attendance was Lew Spencer, our corporate counsel, whose self-appointed role was to keep the rest of us organized. He maintained a three-ring binder with a tabbed page for every idea that could possibly result in a position paper. Next to him sat Travis Marshall, Motorola's chief lobbyist and coordinator of the stable of advisors, lawyers, and lobbyists who represented the company in Washington and advised management on telecommunications policy. Three of those advisors, all lawyers—John Lane, Don Bealer, and Neal Kennedy—were also in the room. So was Len Kolsky, Motorola's regulatory lawyer, who had once worked at the FCC; he was the only other junior staff member in the room besides me. Karl Nygren was also present—he didn't speak often, but when he did it was always wise and insightful, and everyone listened.

Leading the meeting were three of Motorola's key executives—Chairman and CEO Bob Galvin, newly elected Executive Vice President (and soon to be President) Bill Weisz, and John Mitchell, vice president and assistant general manager of the Communications Division. Each had prepared themselves in their particular fashion. Bob, to supplement his methodical morning swim, had walked up the twelve flights of stairs from the basement parking garage—but slowly, so he wouldn't break a sweat. John had also come from his morning swim but, as befit a former water polo player, his workout had been an aggressive regimen testing the limits of his fit but arthritis-ridden body. He privately scoffed at Bob's swim as "hardly worth the effort." Bill, paranoid and competitive, had instructed his driver to take a random route from his home to avoid any potential corporate espionage. For him, winning was not the main thing, it was the only thing. He arrived before the others.

This was our leadership, and they were divided as to how Motorola should proceed regarding AT&T's proposal and impending FCC action.

AT&T proposed using UHF spectrum to build a cellular mobile phone system, made up of low-power transmitter towers that formed radio frequency coverage areas called cells. As callers moved from cell to cell, their call would be handed off from tower to tower. This was different from IMTS, where a single tower blasted a signal over a citywide area, leading to congestion among users. The cell structure, conceptualized twenty years earlier by Bell Labs, was similar to what Motorola had devised and implemented for the Chicago police.

Callers in AT&T's new cell system would be using mobile phones in vehicles: cars, trucks, trains, and so on. To do this at scale, AT&T wanted use of the 900 MHz frequency band, and they wanted a commitment from the FCC to allocate the spectrum if they showed they could build a viable system.[10] The Metroliner demonstration, built by General Electric with cell handoff along the train tracks, was AT&T's way of showing that its system worked as proposed.

AT&T envisioned a cellular system based on mobile phones in vehicles—yet it saw mobile telephony in general as a dead end. The company's internal market research saw little demand for the service.[11] For this reason, AT&T proposed that it and its subsidiary Western Electric be the exclusive supplier of both equipment and service, the same arrangement as in the wired telephone business. If only a small number of people were going to buy mobile phones (a view shared by Commissioner Lee), then why bother opening up the market to other companies? It wouldn't be much of a market anyway.

At Motorola we didn't like the sound of that: AT&T was proposing to extend IMTS (for which Motorola built equipment) but exclude us and others from hardware manufacture. This was far from the worst part.

Because AT&T did not imagine a large enough market for mobile phone service to justify building an expensive network of cell towers, it requested something else. AT&T asked the FCC for authorization to deliver two-way radio communications and a proposed new air-to-ground communications service. In other words, AT&T proposed to be the sole provider of cellular and air-to-ground services and to compete with existing land mobile manufacturers. This proposed arrangement was not only unfair, it was unworkable. The monopoly portion of the business would have guaranteed profits, with which the company could subsidize the competitive portion of its business.

AT&T was seeking to escape the restrictions of the 1956 antitrust consent decree, which prevented the company from entering the land mobile business and restricted its manufacturing to the wired telephone system. Motorola dominated land mobile services, principally the two-way radio business with taxi dispatch, public safety systems, and more. This was Motorola's most profitable area of business. We were a near monopoly with about 80 percent market share supplying the Radio Common Carriers.

And now here was AT&T proposing to enter—and essentially take over—the land mobile business and all air-to-ground communications, and to build a new wireless cellular system. They wanted everything.[12] Worse, the FCC was poised to give it to them.

Five and a half decades earlier, the Bell System monopoly had been born with the Kingsbury Commitment of 1913. That agreement settled an antitrust lawsuit by the Department of Justice against AT&T. Instead of breaking up the telephone company, the Kingsbury Commitment created a government-regulated monopoly. AT&T, led by Theodore Vail, agreed to be regulated and to open up its long-distance lines to other companies—but it argued that telephone service, in the public interest, lent itself to a "natural" monopoly and that only AT&T was in a position to build a national communications system. The government agreed.[13]

The Kingsbury Commitment predated the FCC's creation by twenty years, but in the decades since, according to some, a "cozy arrangement" between AT&T and the FCC had served the interests of both.[14] The FCC saw telephone service—and potentially cellular service, too—as a natural monopoly. By coming back to AT&T in 1968 to see if the giant's 1958 spectrum request still stood, the FCC had signaled that it saw AT&T as the only company that could provide mobile phone service. Extending AT&T's monopoly was the comfortable and easy decision: it would let the FCC off the hook from trying to actively regulate a competitive market. Worse, the FCC agreed with AT&T that cellular service wouldn't amount to much and was a secondary concern.

The FCC itself was not, at this time, always the height of technical expertise. During one meeting between commissioners and Motorola, one asked, "What's a megahertz?"[15] The exception to this was Ray Spence, the highly respected chief engineer at the FCC. Ray was often invisible, but he was highly influential; commissioners listened to him. I was struck, on my first visit with him, at his relatively austere office. Though substantial in size, he shared the space with his deputy. Sitting at their ancient but serviceable wooden desks, they faced each other at opposite ends of the office and shared a secretary. Even in Motorola's relatively perk-free environment, the top engineering job in an important division would have warranted a private corner office. In Ray's shared office, there were documents stacked everywhere; chairs had to be cleared if a visitor stayed long enough to talk.

I had grown friendly with Ray as a fellow engineer, and he was candid with me about the direction in which the FCC was headed. He appreciated technology and its impact on spectrum, and he sincerely wanted to be sure the radio spectrum was used in the public interest. He confided that he was not crazy about monopolies being in control of radio spectrum. Despite this bias, the AT&T proposal was very attractive to him.

The FCC saw the cellular approach proposed by AT&T as a major advance in using radio frequency spectrum—which it was. They saw the Bell System as the only entity with the technical and financial resources to implement the technology. Not incidentally, granting AT&T's requests would also relieve the FCC of a major task: dealing with the nonstop requests for new spectrum allocations from a variety of entities like the land mobile industry and Radio Common Carriers. The FCC had just been deluged by thousands of requests, in 1969 and 1970, to build and sell private microwave lines. The agency couldn't deal with the volume—and didn't want to. It simply approved all the "specialized" carriers as a class instead of on a case-by-case basis.[16] That experience likely provided strong temptation for FCC staff and commissioners when it came to cellular. If AT&T's proposed cellular service was as spectrally efficient as they claimed, and if AT&T could actually build the nationwide infrastructure to serve a public wireless (car) telephone market as well as the land mobile and air-to-ground markets, the FCC—and specifically Ray's staff—would be relieved of a huge burden.

AT&T maintained an army of lobbyists in Washington, with a handful assigned to each FCC commissioner and their staff. The company knew how to work the system, because in a lot of ways it had designed and defined the system. Despite their political sway, in an ironic twist, a Bell Labs employee later claimed that Motorola "was politically much cleverer than we were. . . . They were much better at PR."[17]

〜❀〜

It certainly didn't feel, at the time, like we were very good at politics and public relations. In some ways, Motorola made things harder on itself, because, within the company, we weren't as aligned as we could have been.

One day in October 1969, Curt Schultz, Motorola's chief systems engineer, was meeting with the FCC's staff engineers at their headquarters in Washington. Schultz told them in no uncertain terms that it was technically impossible to use the 900 MHz UHF frequencies for two-way communication purposes. Thus, AT&T's proposal to use that frequency band was ridiculous. At the exact same time, I was with Motorola's Washington staff at the International Club a few blocks away, meeting with other FCC staff. Over a lunch of softshell crab and Maine lobster (in those days, we were permitted to buy lunch for government employees), we eloquently described how the 900 MHz band was perfect for two-way radio communications. We were also attempting to convince them that Motorola's engineers were the ideal people to design and build the radio equipment needed to make use of this band.

Why the discrepancy? Many of Motorola's operating executives, especially those in the sales department, were focused on their bottom lines, their profit-and-loss statements. As they should have been. Naturally, their opinions on how to proceed reflected that focus. For them, Motorola's mission was selling two-way radios, and their bonuses depended on succeeding at that mission. This required acquiring more radio spectrum that could support immediate sales growth and selling equipment that was already in their portfolio, not equipment that would take years to develop. That meant putting all Motorola's lobbying weight behind securing access to the 450 MHz band. Publicly, then, their position was that two-way radio communications at 900 MHz was technically impossible.

From a short-term sales perspective, they were right. Mobile telephony, in the form of IMTS car phones, was a tiny part of Motorola's business and didn't appear to be much of a promising opportunity.

I came at the issue from a different perspective, oriented toward the long-term future. I believed that the 900 MHz portion of spectrum would, within a decade, be suitable for land mobile uses; John Mitchell agreed. We saw an opportunity to expand mobile and portable communications—creating new product and service areas for Motorola—in that part of the spectrum. We also knew that parts of Motorola were making progress in developing the technology needed to make two-way radios and telephones work at 900 MHz. Eventually, we thought, Motorola would use the 900 MHz band for our land mobile customers and compete with AT&T for spectrum access. Any FCC action that would exclude us from the 900 MHz band would eliminate the possibility of Motorola's involvement in a technology that, while still only in its very early stages, could change the face of the industry.

Still, much work remained in device and system technology for 900 MHz communications to be practical. The radio-related devices like receiver transistors and power amplifier transistors were not yet in production, and experimental devices were costly and unpredictable. The characteristics of 900 MHz radio waves differed from those at lower frequencies; they were, for example, more easily interfered with by foliage and buildings. Leaders at Motorola who were more focused on the short term and the bottom line were skeptical about taking on AT&T. The engineering and legal costs of an all-out war would be a drain on the company's finances. Their view was that our differing position was unreasonable, and they let it be known privately and publicly.

It was no small matter, as Karl Nygren later recalled, to take on AT&T. Motorola was a supplier to them: "You don't just walk away from a major customer unless it's a major strategy decision." To take on AT&T—a monopoly, the world's largest company, and a customer—would be a "serious decision."[18]

Here was Motorola facing a major corporate and regulatory threat—and we were internally divided. Our discontent and inconsistency broke out into the open during a January 1970 FCC hearing. This hearing, three years before my dramatic tape-ripping performance, hinged on the question of allotting higher spectrum frequencies. Along with Bill Weisz and John Mitchell, I attended and testified at the hearing. A month earlier, I had conducted a practice run of my testimony with some of our industry partners in NABER. It hadn't gone well. When I finished, all I received were vacant stares from spectrum-hungry two-way radio manufacturers—I had stupefied a group of industry insiders who were meant to be our collaborators. How would I persuade the FCC of the value of competition and Motorola's approach if I couldn't even clearly communicate to our allies?

To better prepare, I worked closely with John Lane, one of Motorola's lawyers. He was well connected and street savvy. John went through my prepared remarks— the speech that had all but put our NABER allies to sleep—word by word, deleting most of them, yet without altering the substance. It was a master lesson: to this day, before every speech I deliver, I remove any word or phrase that doesn't add to the meaning or context of my points. My experience with John contributed to my career, though I'm not sure it contributed to our chances with the FCC.

The FCC commissioners made clear in their questioning that they viewed Motorola's arguments as self-serving and confused. In my 1970 testimony, I gritted my teeth and gave the company line that the 900 MHz radio frequency band in which AT&T proposed to operate was unsuitable for land mobile operations. I knew that, in time, we would solve the technical and economic problems that made 900 MHz unsuitable and that the frequency would then become valuable for land mobile. But I also knew that the process would take at least ten years, so, in some sense, I was telling the truth.[19] Yet the FCC knew we were working on that technology and that we had made the case to them that Motorola could do it.

The FCC also called out Motorola for trying to play the role of potential innocent victim by opposing an extension of AT&T monopoly. After all, we made up over 80 percent of the two-way radio business and were arguing for more radio spectrum at 450 MHz. Later commentators have criticized Motorola for this contradiction. Did we want to protect and expand our position in land mobile communications? Of course! Any company would. But that didn't put us in a very strong position to criticize AT&T's attempts to procure the same advantage. We found ourselves defending the position that free enterprise was threatened by AT&T's offer to provide a competitive business in the very frequency band that we had declared unsuitable—all while we attempted to secure additional radio spectrum to continue our dominance of the two-way radio industry. Not the best position to be in.

The very fact that the FCC was holding hearings did reflect some progress driven by Motorola. In the 1950s, during the early battle over microwave communications equipment, Motorola's lobbying in Washington changed the way the FCC dealt with the Bell System. Up to then, the agency had kept watch over the monopoly through what it called "continuing surveillance." All that really meant was "informal discussions" over expensive lunches. Motorola's regulatory actions around microwave equipment caused the FCC to move to "a more formal basis."[20] That meant hearings. There was still no guarantee that Motorola and our allies would prevail in the early 1970s, but our efforts two decades before had made the opportunity possible.

The 1970 hearing kicked our efforts into high gear. Over the next three years, thousands of pages of FCC filings were produced by Motorola, AT&T, and others. Typically, I wrote the technical version of a plea, and Len Kolsky, our regulatory attorney in Washington, translated it into legalese and added the policy issues. Since there were no word processors at that time, a team of typists produced paper documents that were cut into pieces, pasted together, and retyped repeatedly. Finished documents were rushed to O'Hare Airport so they could be delivered to the FCC by the following morning. Fortunately, there were no security lines at the airport; in 1969, there was no security at all. We worked until the last minute, then sped to the airport to drop Len and Travis off ten minutes before takeoff. Somehow, they always made the flight.

Even as we pushed the FCC to reject the AT&T proposal over the years from 1970 to 1973, we couldn't ignore the views of the C&E (Communications and Electronics) Division within Motorola. It was the heart of the company, the engine of our overall growth and profitability. What mattered for C&E at that moment wasn't AT&T's cellular proposal but securing of more spectrum for two-way radio systems.

For the company's long-term growth, however, the strategy was clear to me. Competition was in the public interest, and portability was the future of communications.

In the short term, the C&E folks were right to worry. AT&T had correctly pointed out that current spectrum use by land mobile equipment was inefficient— and Motorola's two-way radio business was the biggest in the land mobile arena. To counter AT&T and maintain our credibility with the FCC, we needed to demonstrate that Motorola could more efficiently use land mobile spectrum and do it at higher frequencies. We needed to demonstrate better technical abilities at higher frequencies of the spectrum.

For years, I had been talking about more efficient land mobile communications with Roy Richardson, our director of research. We knew from our own discussions that AT&T was right about one thing: licensing a single channel to a customer for their exclusive use was inefficient. In its cellular proposal to the FCC, AT&T proposed to allow groups of its customers to share channels. This was known as "trunking," and AT&T, by moving past Motorola on the issue, was giving Ray Spence and the FCC one more reason to approve their far-reaching proposal. To meet the challenge posed by AT&T and meet the short-term strategic needs of C&E, I persuaded Roy to assign some people in his Applied Research Department to work out the details of a trunking system for land mobile communications.

What I described to them was a system that aggregated twenty radio channels and allowed customers to access any one of a group of channels independently. In examining the differences between telephone and land mobile communications, I had observed that the efficiency gains from aggregating channels, trunking, peaked at about twenty channels for dispatch systems, like police and fire departments, and thirty channels for a system serving phone calls. Each customer would treat a radio channel as though it was their own, and there would be no interference with others. For practical purposes, each company had their own private system.

The C&E Division had been operating community repeaters for years, which were radio communications sites equipped with base stations and antennae. These repeaters were leased to customers for use. Community repeaters efficiently provided for calling groups of people (as in the police dispatch for "calling all cars") as well as one-to-one calls through press-to-talk service. Yet use of community repeaters was, by the late 1960s, technically primitive and inefficient, limited to just five channels. And AT&T was now calling Motorola out on this outdated technology. C&E executives didn't want a new technical solution—they just wanted more spectrum for their inefficient technology. Roy's team, by contrast, created a trunked system where many users could share multiple frequencies on a first come, first served basis.

Once the technical solution was figured out for what I envisioned, I had to sell it to Ray Spence. Ray sincerely wanted to be sure the radio spectrum was used to maximal public interest. What Roy Richardson and his team came up with resulted in higher efficiency, better use of spectrum, and improved performance. Ray, as a true engineer, recognized a technically superior solution when he saw one. Based both on the efficiency of the new trunked system and the bond I developed with Ray—through our shared engineering enthusiasm—he approved our solution. The FCC would eventually require that all land mobile systems at 900 MHz with more than five channels use trunking technology.[21]

The trunking system I dreamed up, that Roy and others at Motorola helped make real, was the basis for what became known as Specialized Mobile Radio Service (SMRS). This became the platform for successful businesses such as Nextel.

Trunking solved one issue for the FCC around private land mobile services. Yet there was still resistance to it from within Motorola, from C&E executives who said their customers would never buy something that cost, by my estimates, about 15 percent more than they currently paid (notwithstanding improved efficiency and performance and thus lower costs in the long run). Ten years later, the SMRS business represented 80 percent of C&E's two-way radio revenue.

More importantly, trunking proved two things to the FCC. First, that land mobile service could remain separate from cellular telephony and be more efficient. Second, that competition was crucial for innovation. In its order requiring trunked systems, the FCC explicitly opened the door to SMRS entrepreneurs who were trying to exploit underutilized spectrum.

⌒⚹⌒

By the autumn of 1972, rumors were floating around Washington that an FCC decision on the fate of radio spectrum was imminent. Motorola's corporate paranoia had reached fever pitch. Even as the company, mainly through Curt Schultz, continued to say that 900 MHz was unsuitable for land mobile communications, Motorola had demonstrated increasing technical expertise on the use of 900 MHz frequencies and through trunking systems. Curt was not part of that work, and by the spring of 1973 Motorola had changed its public position. Our internal conflict and change of tune had not escaped the attention of FCC commissioners and staff.

Commissioner Robert E. Lee—he of the "frivolous" spectrum use view—had pin-on buttons distributed at the May 1973 hearing that asked, "WHERE'S CURT SCHULTZ?" Now that we had changed our position, the FCC doubted Motorola's capacity to deliver at 900 MHz.

We were boxing ourselves in and needed a way out. It was at that May 1973 FCC hearing that I would get everyone's attention with my tape tearing. But to truly get the FCC's attention, settle their doubts, and shape their decision making, we needed something big.

John Mitchell had been thinking the same thing, and, sometime in 1972, he proclaimed that Motorola had to do something "dazzling." He assigned Jack Germain, who ran the Mobile Products Division, to prepare a dazzling demonstration. Jack reached into his engineering departments and instructed his team to stop all product development, if necessary, and to use all their resources to assemble a complete model line of mobile and portable two-way radio equipment operating at 900 MHz. We had to prove to the FCC that we could follow through, technologically, on the reversal of our position on the suitability of 900 MHz frequencies for land mobile. At the same time, John Mitchell engaged Jerry Orloff, head of public relations, to work with Jack to prepare an impressive extravaganza to be held in Washington, suitable to impress the FCC and members of Congress. Travis Marshall, our chief lobbyist, was assigned to generate anticipation among these prospective audiences.

I was crestfallen. How could there be a big show without my involvement so I could tell my side of the story? Didn't John remember what a good demo team we were, including our IMTS success? I tried to imagine a riveting land mobile radio demonstration but came up short. As easily as I communicated my passion inside Motorola, for years I had had trouble exciting the general public or politicians about the importance of the land mobile radio business—even my mother, who, as proud of me as she was, had no idea what I did for a living. She certainly would rather have been able to brag about her son, the doctor; even a lawyer would have been better than a radio engineer. The reality was that, to outsiders, land mobile was boring. Sure, the FCC staff knew about our important contribution to productivity in many workplaces, and to public safety in police and fire departments, but we were trying to get the attention of Congress and other politicians, an entirely different audience.

To preserve the viability of Motorola—and to save the future of communications from the chokehold of a monopoly—we needed to do something *truly* dazzling. Something far beyond a basic land mobile demonstration. What could we possibly do to stave off AT&T's takeover of cellular?

Then it hit me. We needed to make a phone call.

THE THEATER OF
INNOVATION

I 'm coming to the introduction of the first cell phone. First, let's talk a bit more about showmanship, a skill I learned early from my mother, as I watched her selling corsets in downtown Chicago.

The people who hold the purse strings in any organization (like Motorola) are inherently conservative, and we're fortunate that they are. Someone must sort out the unending requests to fund the necessary costs of product development and introduction. In response, engineers discovered that, if we're ever going to have a chance of bringing a product to market, we need to be able to sell our project internally. We needed to be showmen.

Showmanship is a necessary ingredient in any sales pitch. There is much lip service given to the need for creativity, for new ideas, for new and better ways of doing everything, but when the time arrives to deliver a check, projects are more likely to be rejected than accepted. New ways of doing things involve risk. Unless the value of a potential change is great enough to justify the risk, the change will be rejected.

Expressing value is crucial and involves all the tools of selling and showmanship. Presentation is crucial, and I was blessed by having that drilled into me early in my career.

In our IMTS presentations, John Mitchell and I spent a fair amount of time behind our magician's table, making presentations and (hopefully) beguiling audiences with our new ideas. Don Linder called us "Amos 'n' Andy." When I spoke to him for this book, he reminded me that John and I "finished each other's sentences and each fed off the energy of the other."[1] This is how persuasion and selling work. You need a fair amount of theater around it. This is important externally and internally.

In the spring of 1965, the management contingent of the Motorola Communications Division was enjoying a festive evening at the bar of the Valley Ho Hotel in Scottsdale, Arizona. We had spent the day negotiating with our counterparts in the Semiconductor Division and had pulled off a coup. The meeting was crucial for the Communications Division. We were pleading with the Semiconductor Division to design and manufacture a unique radio frequency transistor that would assure continuation of our competitive superiority against GE and RCA. These industrial behemoths were huge compared to Motorola, but we were, by far, the leader in technology and market share.

The semiconductor managers were sympathetic, but at our meeting two days before, they explained that we didn't buy enough devices from them to make our business attractive compared to their external consumer manufacturing customers. A transistor design for a car radio would generate the sale of millions of transistors; the best we could do was tens of thousands. The only way the Semiconductor Division could meet our needs, they said, was if we could come up with a *really* big order.

Our team was undeterred. There was no way we would leave Phoenix without a favorable commitment by the semiconductor folks to design and build a custom transistor that we needed for a two-way radio breakthrough. At that day's meeting, the Communications Division was represented by Bill Weisz, general manager of the division; John Mitchell, his number two; and me and Jack Germain, chief engineers. We were making our case to a team headed by the assistant general manager of the Semiconductor Division. He was an erudite and cultured gourmet in contrast to the down-to-earth communications contingent, whose idea of gourmet was a sausage and pepperoni pizza garnished with olives and mushrooms, smothered in chili sauce. I had dined with him several times in Chicago and enjoyed his company. He had only recently joined the Semiconductor Division. When the company moved him to Phoenix, he had insisted that his wine collection be moved with him, at company expense, in a temperature-controlled truck. We had to get him to balance the benefit to his division and the corporation as a whole. After all, his stock options in Motorola, which he depended on to secure a lavish retirement, were valued by the performance of the corporation as a whole.

In his inimitable way, John saved the day. He began, "You told us that you needed a really big order from us. Well, we listened to you."

With a flourish he opened the conference room door. Two of our staff entered carrying a four-by-six-foot sheet of plywood painted white with details replicating a supersized production order for a substantial number of transistors. We were delivering a BIG order.

When the laughter subsided, we explained that we were making a multiyear commitment that we knew we could fulfill. Whether it was our humor or sincerity, the tenor of the meeting changed. The semiconductor team understood that if they turned us down, they would get pressure from the chairman of the board all the way down. They agreed to design and manufacture our transistor.

Several years later, in 1971, Motorola announced the Pageboy II pager along with several versions of the paging terminal at a series of customer meetings starting in Phoenix. The audience was comprised of the Motorola sales force as well as key customers including Homer Harris, whose daughter, Arlene, I married years later. The dramatic high of the presentation used the opening scene of Stanley Kubrick's *2001: A Space Odyssey*, which had been released three years before. A caveman hurls a bone into the air. The bone tumbles into the distance and reappears as a spaceship, still tumbling. All the while we are listening to "The Sunrise" from the Richard Strauss tone poem *Also sprach Zarathustra*.

Except that, in our case, the bone reappeared not as a spaceship but a tumbling Pageboy II pager. Audiences roared with appreciation, even the salespeople, who had been impatiently waiting a year or longer for the delayed product introduction. The agonizing series of production delays and broken promises was forgotten.

For my announcement speech, our public relations team rented a magician's glass cage in which I suddenly appeared out of thin air. Despite my love of a show, I had told them I preferred to be introduced as just a guy who was an engineer. But they overruled me—they wanted to emphasize the magic of creating this pager and its supporting systems. They were right. Elements of showmanship were needed. Just as we needed some theater to introduce the first cell phone.

THE BRICK CHRISTENED DYNATAC

Birth of the Cell Phone

"We have been very careful, up to the present time, not to state to the public in any way, through the press or in any of our talks, the idea that the Bell System desires to monopolize . . . but the fact remains that it is a telephone job, that we are telephone people, that we can do it better than anybody else, and it seems to me that the clear, logical conclusion that must be reached is that, sooner or later, in one form or another, we have got to do the job."[1]

— A. H. Griswold, *AT&T executive*

That statement sums up the Bell System's outlook when it came to cellular telephony in the early 1970s—feigned reluctance, strident modesty about serving the public interest, and barely concealed arrogance. Actually, this view captures AT&T's attitude during most of the twentieth century toward technologies of all types. This statement was made in 1923, when AT&T was already an established and government-sanctioned monopoly. It was made in regard to AT&T seeking dominance in radio broadcasting. Yes, that's right—broadcasting of shows and programs over the radio. The company controlled the long-distance network, or the telephone infrastructure, so radio was merely an extension of their monopoly. That was the natural order of things—to Bell's way of thinking, at least.

Radio broadcasting dominance was turned over to the Radio Corporation of America. But the Bell System attitude framed its approach to cellular telephones and the radio spectrum in 1972: *If somebody has to do this (for the public, of course), it may as well be us, because we're the Phone Company. But we need total control to make it work.*

In 1973, the FCC was about to give it to them—unless Motorola could change their minds. No amount of dramatic tape ripping at a hearing was going to be enough.

We needed more—way more—than the two-way radio demonstration proposed by John Mitchell. Maybe, just maybe, that would prompt the FCC to spare a thought for Motorola and others and leave a little room for competition in the land mobile business. But AT&T's proposed monopoly in cellular, two-way communications, *and* air-to-ground communications was on the way to FCC approval. The FCC had asked AT&T for a proposal; AT&T had submitted one. In the late summer of 1972, we were hearing through back channels that a decision was near and it wouldn't be friendly to Motorola and our allies.

We needed something that would show the FCC what the future of telecommunications, in a competitive marketplace, could really look like. Our upcoming testimony in the spring of 1973—where I would pick apart and argue against AT&T's claims (with a little dramatic flair)—would need to be supported by a monumental display of technical prowess.

We needed to actually show, not just tell, the FCC that AT&T's claim that only it, as a monopoly, could deliver cellular service, was baloney. What if Motorola could demonstrate a vision of the future of *real* mobile telephone technology? I knew such a demonstration could be done. I knew such a demonstration *had* to be done. "We have to do something spectacular," I argued in an internal memo.

⌒※⌒

I couldn't get the vision of portability out of my head. The work on pagers and police radios had been pushing Motorola in this direction for years. Our present work on telephones was wrapped up with IMTS, the car phone service with AT&T. When it came to the future of telephones, they had to be portable, right? Which meant not chained to a vehicle. I'd been roaming around the various labs of our research departments for several years—well before the Bell cellular proposal to the FCC—encouraging engineers to work on devices and circuits suitable for future generations. So I knew that Motorola had created all the technological components of a portable telephone system that would be superior to the Bell System's offering—or, at least, what we knew about it.

In late autumn 1972, with less than six months to go before the hearing, I barged into John Mitchell's office: "John, I understand how important it is to demonstrate two-way radios at 900 MHz. But if we're going to impress the FCC and the politicians, we need to do something a lot more glamorous than two-way radios. We can have a prototype handheld portable telephone built in time for the Washington demonstration. Why don't we out-Bell Bell?"

He thought for a moment. "I think you're right, Mart, but you'll need to get it done for a demo in April. And I won't count on it until you can prove to me it'll be done on time."

As usual, he was right. "The Motorola Pageboy introduction was eighteen months late," he reminded me. "The AT&T threat is too important to allow even a few days' delay."

I didn't have eighteen months. The end of 1972 was upon us, and the next round of FCC hearings was scheduled for the spring. Jack Germain's group was already putting together the two-way radio demonstration to be given to the FCC in Washington.

I had three months. Complicating matters, I had no independent resources at my disposal. At this time, I was vice president and director of systems operations in the Communications Division. But the puzzle pieces needed for a truly portable phone were scattered throughout Motorola. As far as I could tell, I was one of the few people who knew (from my roaming ways) that all the necessary technologies were available or being worked on already. That was my leverage. I would have to cajole and persuade and inspire. To get started, at the beginning of December 1972, I paid visits to three people.

The first was Rudy Krolopp, head of the industrial design group; his team did the industrial design for all of Motorola's communications equipment. Rudy oversaw aesthetics and user interfaces. We needed an attractive device, and Rudy was a master at industrial design. He was independent, opinionated, and stubborn—as well as charming and imaginative enough to get away with those attributes. Better yet, he was a genius at designing equipment that was attractive while getting the job done.

"Rudy, I need you to design a handheld wireless telephone, a portable cell phone."

"What the hell's a portable cell phone?" Rudy responded.

I said, "Well, you know, it's a telephone you can carry with you all the time, and it's got to be really jazzy because we're going to use this for a demonstration."

Rudy recalls that I picked up his office phone and said, "If I took scissors and cut the cord off, and I walked around and had to do everything with this phone without the cord, that would be a portable cell phone."

We'd had conversations like this before, and sometimes they resulted in commercial products. Sometimes they resulted in ideas that were too futuristic or downright crazy to bother with, including a wristwatch pager that was never produced but that anticipated the cellular watch.

I continued, "We are taking on the biggest company in the world and showing them that we know how to make a portable telephone that can call anyone anywhere in the world—and that we can do it better than they can."

Rudy thought about it for a minute. "This sounds like it could be great fun, and my guys will be really turned on by the project. What are the specs and when do you need a concept?"

"No time for concepts, Rudy. We need to get this done by March, so we need a model in a few weeks."

When I interviewed Rudy for this book, he told me, "I didn't say it, but I was thinking, 'Are you out of your mind?' But I never said no to you."[2]

For another hour, I described the handheld portable phone vision, the audience for our demonstration, and the reasons why the demonstration was so important for our business. Rudy got it all: "I'll do it."

Rudy and the industrial design department did not report to me. After our conversation, he stopped working for all his internal Motorola clients, but he didn't stop charging them for his services. I remembered my lessons from John about budgetary allocations, the differences between what was on paper and what a passionately persuasive person might arrange. Checking on Rudy's progress was a challenge. If I wandered into his workshop too often, he would feel that I was pushing too hard. If I didn't show up frequently enough, I wasn't expressing adequate interest. I was finding the balance John had told me about, knowing when someone was close to the edge of the cliff and when to offer the cup of coffee. In Rudy's case, every three days turned out to be about right. My role, as described by another colleague, was "keeping us going, trying to keep our morale up."[3]

The third time I visited him, Rudy said, "We're getting close. I decided that this is important enough to come up with more than one model, so I put my five best designers on the project, and they've been doing this day and night. You're still crazy, but we might be able to meet your ridiculous schedule."

Two weeks later, I arranged for a dinner in a private room at Lancer's, the restaurant across the highway from our offices. In the middle of a large banquet room (large enough to seat one hundred people), seven of us sat around a single

Ahead of its time: flip mouthpiece design concept produced by Rudy Krolopp's team for the Motorola DynaTAC, 1972.

Double flip design concept produced by Rudy Krolopp's team for the Motorola DynaTAC, 1972.

table. The designers were itching to show the results of their work to me and Rudy—and each other. A little competition had spiced up the process. Despite the demanding schedule I had given the team, I expected to see renderings, drawings of design ideas. I got a lot more than drawings.

Each designer stood and eloquently presented a physical model of his vision. Their pride lit the room. They explained the functionality of the features as well as the aesthetics. Each model was beautiful—and they were all predictive of the future. There was a slider and a flip phone. One that folded top and bottom like a book. One that looked a bit like a boot. Each was the size of a cell phone today.

Retractable design concept produced by Rudy Krolopp's team for the Motorola DynaTAC, 1972.

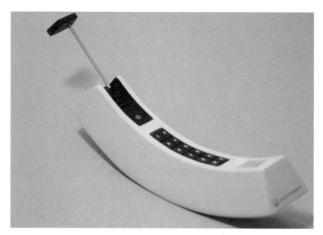

"Banana" design concept produced by Rudy Krolopp's team for the Motorola DynaTAC, 1972.

We needed to decide. The design we chose had to be turned into a working phone in just a few months. We talked about the objectives and timing. Rudy had clearly influenced each of the designs, and we made sure each designer had an opportunity to express his views, so each knew they were part of the ultimate decision. (All the designers at the time were men.)

In less than an hour, we had consensus: Ken Larson's design. While not the most creative, the boot—or "shoe phone"—was a single block, attractive and original, with a conceptual relationship to a classic telephone. That simplicity is what tipped the scales. We wanted to make sure to minimize the number of things that could go wrong in the demo. It had to work.

The "shoe" or "boot" phone design produced by Ken Larson, which became the prototype for the Motorola DynaTAC, 1972.

The design decision was the easy part. Ken's "shoe phone" model had to be made into an actual phone, with all the requisite parts inside it.

This had never been done before. I knew that all of the individual technologies that were required to make a handheld portable cell phone were at various stages of development in our research departments. We had been working for a few years, for example, on using an integrated circuit, a semiconductor chip, to replace sections of a two-way radio that were previously built with hundreds of individual components. Adapting those to handheld phones was only a few steps away. But it wouldn't be easy.

When I envisioned a user of this new portable cell phone, I thought of someone talking on it while they were on the move. That seemingly simple image required several breakthroughs. It needed to be full duplex: callers needed to be able to talk and listen at the same time. It wasn't merely an extension of two-way radio service. To do this we'd need a radio frequency power amplifier that could produce a watt of energy continuously and reliably at 900 MHz. Such a device did not yet exist; we were still struggling to make 450 MHz amplifiers work reliably. The radio receiver needed to be extremely sensitive to extract the impossibly weak 900 MHz signal from the air. And we needed a tri-selector, the device that made full duplex possible in the IMTS radiotelephone, except that our tri-selector needed to be miniature. The IMTS version was almost as big as our entire radio. A miniature tri-selector had never been built. And the IMTS tri-selectors operated at lower frequencies; we had no experience building one for 900 MHz.

The new phone also needed to use hundreds of mobile radio channels, but we had never built a two-way radio with more than a half dozen channels. The frequency synthesizer that could tune the radio to one of hundreds of radio channels did not commercially exist, although I knew Chuck Lynk, one of our engineering managers, had a design for one in his head. It could be built if the Semiconductor Division of Motorola would give us one of their experimental computer chips. This was 1972; there were no large-scale integrated circuits. We would need to use a large number of small-scale ones. Fortunately, our engineers had already been working on designing a new generation of integrated circuits. That work was still developmental but, hopefully, they could rush something out of the lab.[4]

All of these requirements were breakthroughs in their own right. But they had to be put into a battery-powered, portable handset using hundreds of hand-wired individual components that we couldn't yet put on a single chip. Many of the features had yet to be tested—but they had to work, or the entire project would be a failure.

These were just the deal breakers we knew about. We had no idea what other pitfalls might lurk ahead of us.

After Rudy was off and running with the project, the second person I approached for my core team was Roy Richardson, our research director, whom I was already in constant touch with about trunking. This time, I said, "Roy, we're going to do a crash project, and I need your help."

"What's the project?" he asked.

I laid it out as plainly as I could. "We're about to do a demonstration of Motorola's technology in Washington. John Mitchell thinks we need to persuade the FCC that we are expert at building 900 MHz equipment. Jack Germain is building a whole bunch of mobiles and portables to demonstrate to the FCC. I think that approach is really boring. Instead, we're going to wow the politicians in DC with a portable, handheld cellular telephone, and we're going to do that three months from now."

Roy considered me for a moment, probably trying to determine if I was pulling his leg. "That's impossible! I can tell you why it's impossible, but I know you won't listen."

"You're right," I replied. "For anybody else, it's impossible. But you have the technology and the people to do almost everything a portable design needs, and if there's anything missing, we can get it from somewhere else inside the company. You can have whatever resources we can dredge up from any part of Motorola. Just tell me what you need and who you want to assign to it."

Roy proceeded cautiously, but he was warming to the idea. "Well, Don Linder can build the radio transmitter and receiver, if we can get someone to make a 900 MHz power amplifier. Don has never built a complete portable, but he's the best we have. I think I can assign Linder and a few of his engineers."

I kept pushing. "I think you're going to assign everybody in your department who can contribute anything. This is more important than anything else the company is doing. We've got to meet the schedule. Failure is not an option." It was, Chuck Lynk later recalled, "unprecedented" at Motorola to devote so much staff and resources to one project with such a "short time constraint."[5]

Roy shook his head but went with me to visit the third person I wanted to join our effort. Don Linder was in his laboratory. As usual, Don was at his bench with eyes glued to a spectrum analyzer. I greeted Don with an enthusiastic hug.

"What do you want, Marty?" he asked suspiciously. This was not the first time I had approached Don with an off-the-wall request, initiated with a hug. I used hugs with very few people and still reserve the gesture for those with whom I feel a special affinity. I don't know anyone else who does that. For me it's a sincere gesture, so I've always hoped my huggees regard it positively. No one has ever responded poorly to one of my hugs—yet.

The task I gave Don was to combine existing technical features of Motorola products into a single portable device. Even though the individual technical pieces existed or were under development, they had never before been put together in one device.

Don was incredulous at the timeline, and even more so when I showed him Ken Larson's three-inch-tall model.

"There's no possible way we could ever put anything useful into that tiny amount of space," Don said. "The battery alone will need to be three or four times as big as the entire model."

This was exactly the response I expected. John Mitchell had taught me about what he called "bicycle pumping." As he described it: "When our genius engineers are given a reach assignment, they pump it up as big as they can and make it sound impossible. Then, they extract kudos when they do the impossible. But they are good enough to come through with the goods, and we have to play the game. We have the best engineers in the business; they're entitled to an extra pat on the back."

I told Don, "No problem. You can make it as big as you need to, just so it's handheld and the shape is close to Ken's model. This is a demonstration, not a refined product. And you can have all the people you need to make it happen."

Don Linder in 2007.

"Well, I don't see how we can do it in only three months," he grumbled. "But I'll do my best. You guys better come through, because I need all kinds of help. We've never built a radio with more than six channels, and you want hundreds of channels. We've never built a radio at 900 MHz, and it's a duplex radio on top of that. Impossible. The battery is going to be huge. I don't even have an antenna that will work at 900 MHz."

In other words, Don was hooked—he had gone from impossible to enthusiastic, and he was bicycle pumping like crazy. His excitement built as he explained how he would solve each of the seemingly insurmountable technical challenges.

"Chuck Lynk has been working on a synthesizer that can tune a radio to more than a dozen channels. Going up to hundreds of channels is just a detail. Roy, tell Chuck he's on the team." Now Don was on a roll. "The portable products department engineers demonstrated a 900 MHz portable to me last week; they must have an antenna that will work. I'll get them on the case. The filter designers built a duplexer for the IMTS radiotelephone, which was as big as this entire handheld phone. But they've had several years to miniaturize it and make it a tri-selector. I'll nudge them."

Don had talked himself into accepting the assignment. "I can't guarantee that we'll get it done," he reported, "but I'll do my best."

After Don was on board, Roy Richardson called several other people into his office. Rudy put a blue cloth over the Larson model. I talked about the idea of a portable, handheld phone and how we were going to build it in just a few months. Someone asked, "What's it gonna look like?" Rudy took the cloth off, and jaws dropped around the room.

"Anyone who doesn't believe this can be done," I said, "leave the room."

Nobody left.[6]

By the time Don's team finished, he had assembled several dozen engineers and scientists from his and other labs. As with Rudy's team, I stopped in to visit Don every three days or so. His lab was a zoo. People were working ten- and twelve-hour days. The cell phone was being developed throughout the lab with different engineers handling different parts. John Mitchell was skeptical about whether the project would be finished on time, but he knew how important it was. He visited Linder's lab, too, adding considerably to the motivation of the team.

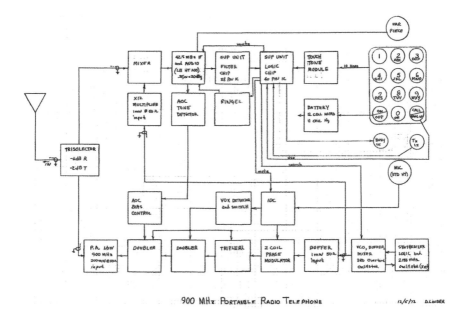

Don Linder's original sketch, from 1972, of what would become the DynaTAC.

Simultaneously, other engineering groups in my division created base stations that would operate at 900 MHz and adapted a telephone switch designed for the existing mobile telephone service to connect the new handheld phones to the AT&T telephone network. The portable had to be able to call anywhere in the world that could be reached by wired phones on the AT&T network.

This was a symphony of great complexity. I was the conductor and needed to be attentive to the sounds and rhythms of the orchestra. I couldn't even think about the possibility that the challenge was insurmountable. I wasn't as expert in any single area as my team members, but they taught me their specialties and I integrated them into the vision. Dick Dronsuth was an expert in power amplifiers; Al Leitich managed the complex audio circuitry; Jim Mikulski knew more about cellular reuse patterns than anyone. The vision materialized with insights from Roy Richardson. John Mitchell, overcoming his early reluctance, became an inspiring champion.

Innovations are ideas that are useful to society and can be demonstrated to be realizable. Inventors, those who are often listed first on patent filings, get a lot of the credit that should go to the specialists who create the basic elements of the inventions. I firmly believe there has never been an invention that did not build upon and rely upon previous inventions.

When I talked with Chuck Lynk for this book, he told me, "We all thought you were crazy." Those of us working directly on it—me and Roy and Don and his team—believed we would succeed. Few others in Motorola thought we could meet the timeline. Despite that lack of belief, I persisted, and with John's endorsement, the team got the support they needed to succeed. Between Jack Germain's two-way radios and our portable, most of the other engineering processes and projects within the division were halted. Whatever we needed—anything—we got.

In the middle of March 1973, I was busy preparing my testimony for the upcoming FCC hearing while also trying to keep tabs on our crash project. Don Linder called me into his lab. The original "shoe" prototype designed by Ken Larson had been relatively small, comparable even to today's cell phones. Because Don's team had to stuff it with about a thousand parts to meet my specifications, the phone grew and grew. We knew the original model would need to stretch to give the electrical engineers room for their parts, but I had no idea just how much stretching would be needed.

Larson's original model had been a half-inch wide, a few inches deep, several inches long, and incredibly light. Now, after being made to accommodate all the necessary parts, Don showed me a phone that was ten inches long, three inches wide, weighed two and a half pounds, and had a six-inch antenna atop it. That was still far lighter than AT&T's thirty-pound car phone system—manufactured, admittedly, by Motorola—that they wanted to put in more vehicles.

The "shoe phone" now came to be called "the brick" because of its size and shape. It could access nearly four hundred radio channels, and the battery could handle about a dozen three-minute calls, lasting up to twelve hours in standby mode.[7]

In just three months, dedicated individuals and teams had created not only a new communications device but also the system of switching equipment and radio stations that, together, would transform the world. The patent that we ultimately secured from the US government—for integration of new and different components and ideas—has Martin Cooper as the first listed inventor. But my idea would have been worthless without the expertise and experience that

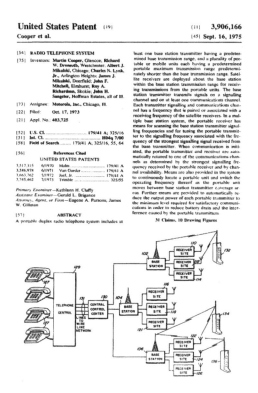

Motorola's patent for a "Radio Telephone System," filed October 17, 1973 and granted September 16, 1975. This was the DynaTAC system patent for a wireless telephone system accommodating truly handheld, portable phones. United States Patent and Trademark Office 3,906,166.

made it real. We had integrated various pieces and components into an entirely new system, not just a product. I came up with the name DynaTAC for our new phone, an abbreviation of DYNamic Adaptive Total Area Coverage, to describe Motorola's approach to cellular technology.

Roy Richardson later told me, "Marty, you persuaded us to do things we never imagined we could accomplish."

By the third week of March, after assembling a new version of the phone with tighter packing, we had a complete unit ready to show John Mitchell. Though he had been keeping tabs on our work, John had never seen the complete unit. He was entranced. He immediately called Jerry Orloff, head of public relations for the Communications Division. We had a demonstration to plan.

On the morning of April 3, 1973, I was awakened by a phone call.

"Bad news, Marty!" It was Rebecca Woodward, a senior member of Jerry's public relations team. "We've been bumped from the *Morning News*."

We were in New York City: me, John Mitchell, Jerry's PR team, and engineers from Don Linder's team and other groups. For the past week, we'd been at the New York Hilton in Midtown preparing and rehearsing. What was originally planned to be a technical demonstration in Washington had escalated into a major PR campaign. Orloff had persuaded management that a demonstration in New York would be most effective in getting the attention of the world press, so New York was added to the agenda. Motorola's field engineering teams, part of the sales division, were tasked with finding rooftops and towers strategically located to best show the freedom that the DynaTAC phone provided.

The PR team had booked the Hilton's penthouse suite, normally reserved for celebrities. Split over two floors at the top of the hotel, the five-bedroom suite featured a grand piano that Mick Jagger played during a 1965 gig as well as the sort of freestanding spiral staircase you'd expect Scarlett O'Hara to walk down. While we could never get the hotel to confirm it, many of us also understood that a certain Hollywood couple—Elizabeth Taylor and Richard Burton—set up house in the suite whenever staying in Manhattan. The engineers used the two-story penthouse as their laboratory for maintaining the two DynaTAC portables, the only ones in existence. This was showtime, and the engineers knew it. They were testing and fixing problems well into the night. Chuck Lynk's memory of that evening is fading, but not the part about him sleeping in a bed in which Elizabeth Taylor had slept.

Downstairs at the hotel, we had reserved a room for the major event of the day, an afternoon press conference. This was it. We had been rehearsing for the past week, trying to ensure that our new cell phone would work as expected. Before the press conference, I was scheduled to be on the CBS *Morning News* to draw attention to the event. We were here to tell our story to the world. Now Rebecca was informing me that I wouldn't be on the *Morning News*.

"I know you're disappointed," she said. "But I called around and got us an interview with a reporter from a local radio station in two hours. It's not the same as the *Morning News* but better than nothing." I knew how disappointed Rebecca was. She was a dynamo who took her work very seriously and had worked tirelessly to get the *Morning News* slot.

I replied, "I'm cool, Rebecca. The press conference is our big event today. But I promise you I'll do a good job with this reporter. Do me a favor, though. Let's do the interview out on the street so the reporter gets the feel of the power of being able to communicate on the move." We'd already made plenty of test calls between engineers. But today would be the first time a public call was made with the DynaTAC.

"Great," she replied. "He'll meet you in the lobby at nine."

I met the reporter in front of the Hilton, and we strolled south on Sixth Avenue while I told him about Motorola's breakthrough. In my hand was the DynaTAC, the grandmother of all the tens of billions of cell phones that have been manufactured since then.

As I spoke, my mind raced, thinking about whom I should dial for the first public call. John Mitchell? My mother? My office? Then I had a serendipitous thought.

I pulled my printed phone book out of my pocket and found the number of Dr. Joel Engel, the engineer running AT&T's cellular program. In the engineering battle that was underway, Joel was my counterpart at AT&T. Together with Richard Frenkiel, Joel led a team of about two hundred engineers focused on building the vehicle-based cellular system. Joel, a PhD with years of experience at Bell Labs, had worked on a variety of projects including one that involved the Apollo moon project.[8] His view, which he expressed publicly, was that Motorola was an annoyance and a hindrance to the process of making cellular service available. We were natural competitors, so I was more than ready to give him the needle.

I dialed Joel's number on the DynaTAC. The phone sent the call instructions through the air—via radio waves, of course—to an antenna on the roof of the Allianz Capital building (which was the Burlington building at the time), across the street from the Hilton. The cellular station connected to the antenna relayed electronic signals to the landline telephone network, which then routed them to Joel's desk in New Jersey. Amazingly he, not his secretary, answered.

"Hello?"

"Hi, Joel, it's Marty Cooper."

"Hi, Marty," he said hesitantly.

"Joel, I'm calling you on a cell phone. But a *real* cell phone, a personal, portable, handheld cell phone."

It was the best I could come up with at that moment. There was silence on the line; perhaps he was gritting his teeth. I don't recall his response other than he was polite and we ended the call after a few pleasantries. Today, Joel says he doesn't remember the call, although he doesn't dispute that it happened. I guess I don't blame him. He has acquired the distinction, among many, as the guy who answered Marty's call—the first public call on a portable cell phone.

I explained to the reporter what had happened on this phone call. Once I pushed the "on-hook" button—the same as picking up a handset on a traditional phone—the DynaTAC sent a series of digital signals to choose the cell site with the strongest signal. Then a dedicated computer connected to both sites; it was called a switching terminal, which needed to tell the DynaTAC's experimental synthesizer which one of hundreds of available radio channels to use. If this synthesizer

An artist's rendering of me making the first public call on a cellular phone,
April 3, 1973, on the streets of New York City.

The World's First Handheld Cellular Phone

The **DynaTAC** by Motorola, Inc.

*Conceived by Motorola Vice President **Martin Cooper***
November 1972
Demonstrated publicly
April 3, 1973 New York City, New York, USA

United States Patent Number 3,906,166
Weight 42 ounces, 1.2 kilogram
Over 400 radio frequency channels in the 900MHz band
Battery life, 20 to 30 minutes usable time

***

DynaTAC, **Dyn**amic **A**daptive **T**otal **A**rea **C**overage

named by Cooper, describing his vision

reliable personal communications adapting to user traffic levels and environmental conditions

***

DynaTAC inspired by **John F. Mitchell**, Cooper's mentor for 25 years

DynaTAC designed and developed by many Motorola leaders:

Industrial Design Team Head, **Rudolph Krolopp**
Designer, **Kenneth Larson**
Applied Research Team Head, **Donald Linder**
Applied Research Department Head, **Roy Richardson**

Contributions made by numerous Motorola engineers and others including:

Charles N. Lynk	Albert J. Leitich	Michael Homa
James J. Mikulski	Ronald Cieslak	William Dumke
Richard W. Dronsuth	John H. Sangster	Bruce Eastmond
Richard Adlhoch	James Durante	Al Davidson
David Gunn	Maynard McGhay	William Rapshys
Merle Gilmore	Gene Hodges	George Opas
Robert Wegner	Robert Paul	Daniel Brown

Edition   Date

_____   _____

A placard that the DynaTAC team created to accompany the original
models produced by Motorola.

did its job, I would hear the dial tone that told me I had a two-way channel into the cell site. I could then enter a number on the DynaTAC's touchtone keypad (rotary dial phones were still common in 1973) so the phone's circuitry could send an audible signal corresponding to the number I dialed. That signal would then travel over a radio wave, exactly as it would have over a wire in a normal call, to reach our cell site. Our switching terminal then introduced the call into Bell's worldwide network so that I could call *anyone, anywhere.*

Suddenly, I felt the reporter abruptly jerk the back of my jacket. I had nearly stepped off the curb into oncoming traffic. We wanted publicity, but that would have been too much.

We still had the press conference that afternoon.

Turnout was disappointing. There were only about a dozen reporters in the room, but I barely noticed. I was still on cloud nine from my sidewalk call. Kicking off the press conference, I tried to call my office—and dialed a wrong number. "Our new phone can't eliminate that, unfortunately," I said, attempting to sound graceful.

One reporter asked whether the DynaTAC would reach Australia. "Of course," I said. "Try it, but I'm sure you know it's the middle of the night there." She called and awakened her mother, astounding everyone in the room. Other reporters used the two DynaTAC phones to make local, long-distance, and international calls. Many asked rhetorically, "Guess where I'm calling you from!" and, "Guess what I'm calling you on!"

After the press conference, we invited everyone upstairs to the penthouse to eat, drink, and continue trying out the phone. Chuck Lynk spent the entire time showing reporters how to push the off-hook button to hang up the line—and then another button to turn off the DynaTAC before its fourteen-volt NiCad battery went dead.[9]

John Mitchell took some reporters and photographers down to the street and made calls on the DynaTAC while posing in front of some pay phones. Newspaper reports the next day noted that passersby were "agape" at a man talking on a wireless phone.

Gee-whiz stories appeared that week in the national newspapers and across the world, all of them favorable. Many cited Dick Tracy's two-way wrist radio as the antecedent for the cell phone. Under the headline "Motorola Introduces Wire-Less Telephone," the *New York Times* noted, "Reception was clear, although the wife of one reporter told her husband, 'Your voice sounds a little tinny . . . There's no res-

information services

MOTOROLA INC.
Communications
Division

1301 Algonquin Road
Schaumburg, Illinois 60172
(312) 358-7900

Mobile FM 2-Way Radio
Portable FM 2-Way Radio
Radio Paging
Communications Control Centers
Visual Communications Systems
Signaling and Control Equipment
Car Telephone
Frequency Components
Microwave Communications
System Parts and Service

Contact: Bob Walz
            Rebecca Knesel

            (312) 358-7900   Chicago
            (212) 685-8154   New York

DYNA·TA·C
Portable Radio
Telephone
System

FOR RELEASE AFTER 11:00 A.M. TUESDAY, APRIL 3, 1973

MOTOROLA DEMONSTRATES PORTABLE TELEPHONE

TO BE AVAILABLE FOR PUBLIC USE BY 1976

New York, April 3, 1973 -- A hand-held, completely portable telephone which will allow its user to place or receive telephone calls from virtually anywhere in a metro area equipped with the new DYNA T·A·C^{TM} system, was demonstrated today by Motorola, Inc.

This new DYNA T·A·C portable radio telephone will operate over radio frequencies and "talk" to any conventional telephone in the world, according to Motorola Vice Presidents John F. Mitchell and Martin Cooper. Mitchell is general manager of the company's communications division, and Cooper is director of systems operations for the division.

"What this means," said Mitchell, "is that in a city where the DYNA T·A·C system is installed, it will be possible to make telephone calls while riding in a taxi,

- more -

Motorola's press release, April 3, 1973, for the DynaTAC.

onance. I knew you weren't calling from a regular phone.'" That kind of complaint, of course, continues today. The *Times* story also noted that the cell phone would be the "answer to telephone addicts' prayers." A prescient article from the Associated Press observed, "There may be no way to escape the strident summons of a telephone in a few years if a portable telephone developed by Motorola, Inc. catches on."

information services   **MOTOROLA INC.**
Communications
Division

1301 Algonquin Road
Schaumburg, Illinois 60172
(312) 358-7900

Contact: Bob Walz
             Rebecca Knesel

Mobile FM 2-Way Radio
Portable FM 2-Way Radio
Radio Paging
Communications Control Centers
Visual Communications Systems
Signaling and Control Equipment

(312) 358-7900 Chicago   Car Telephone
(212) 685-8154 New York   Frequency Components
Microwave Communications
System Parts and Service

FACT SHEET

Motorola DYNA T.A.C TM

Portable Radio Telephone

What is it?

A portable telephone which operates as simply as an ordinary telephone, but
can go wherever a person can go since it operates over ultra-high radio frequencies.

Who will the portable phone user talk to?

Anyone with a conventional telephone anywhere in the world and any other
portable phone user.

Where will portable phone operate?

The unit will operate almost anywhere in a city equipped with the system.
It will work inside buildings, in cars and taxis and on the street. The only places it
won't work will be elevators or sub-basements where radio signals cannot reach.

- more -

Cover page of the DynaTAC fact sheet produced by Motorola, April 3, 1973.

If FCC commissioners, senators, and representatives read the papers, they
heard about the DynaTAC.

A week after the demonstration, *BusinessWeek* ran my picture and called the Dyna-
TAC a "radio telephone" crammed into a "small, hand-held unit." Three months later,
Popular Science featured "New Take-Along Telephones" on the cover, with someone
dialing on a DynaTAC in front of a broken-down car in the background. The article
inside called it "amazing" but, unfortunately, ran in black and white, while an article

about what would soon be the world's tallest building (the CN Tower in Toronto) ran in color. Understandably, for a publication like *Popular Science*, the story focused on the technical functions of the phone. Only at the end did the article nod toward social and cultural implications: "Eventually, the unit could fit your shirt pocket."[10]

At the Hilton, on the day of the press conference and first call, John Mitchell was disappointed at the absence of television coverage. Bill Barber overheard John's complaints. Bill was one of our technicians who had spent the past week climbing up and down buildings to set up and test our cell sites. He was still hovering over equipment in case the DynaTAC calls to the Burlington building from the penthouse went awry.

"I know someone," Bill said. John rolled his eyes in my direction, exasperated at the meddling of a junior engineer. Yet Bill continued.

"I know Bill Leonard, a vice president of news at CBS," Bill said. "He's in my ham radio club." Bill picked up the phone and was soon speaking to this CBS executive. "Bill, I know you like gadgets," Bill Barber said to Bill Leonard. "How would you like to see a portable radio phone?" The answer was: very much. John, Rebecca, and Bill Barber piled into a taxi and headed to CBS. During the ride, Bill suddenly realized that he had no idea if our equipment would work at the CBS office. He started to sweat. He later explained to me, "We'd never tested the phone that far away from our cell sites."

Bill toggled the power button to on as soon the taxi pulled up to the curb. By the time he walked into the building lobby, he had hit the on-hook button—and gasped a sigh of relief. He got a dial tone. Just in time. His friend, the CBS vice president, was walking toward them from the elevator, extending his hand. Bill gave his friend the DynaTAC, and the TV executive dialed his secretary as they boarded the elevator to go up.

"Where did you go, Mr. Leonard?" she asked him, wondering whose office her boss was calling from. "I'm calling you from a *portable* phone," he announced with a grin, right at the moment they all walked up to his secretary's desk. In what seemed like record time, a network reporter from CBS was doing a story on the DynaTAC. They ran a piece on the evening news.

Back at the Hilton penthouse, I chatted with an Australian diplomat. He stopped by to see what all the fuss was about. As we explained and showed him the DynaTAC, the diplomat asked if he could call his wife.

Humans want to communicate—they *need* to communicate—anytime, anywhere, and our goal was to give them the ability to do that.

Me showing a prototype DynaTAC phone to FCC commissioner Benjamin Hooks, 1973.

The demonstrations for political bigwigs in Washington went just as well. Motorola engineers equipped two large vans with equipment for both two-way radio communications and the DynaTAC. Despite the excitement around our cell phone, we still needed to persuade the FCC that Motorola could operate two-way radios in the 900 MHz range. We were still trying to overcome the message conveyed earlier by Curt Schultz that 900 MHz was an impossible frequency for two-way radio.

The vans were set to travel a carefully selected route through Washington, passing through areas in which one of three installed cell stations could reach the DynaTAC. This was an arduous task since the portable phones transmitted very low power that could only reach a mile at the most. Inside, the vans were configured with comfortable couches to accommodate the Washington decision makers.

They were bored by the two-way radios—but thrilled by the DynaTAC. Politicians weren't strangers to mobile phones; they knew all about them, actually. Many had used their influence to acquire car phones, despite their poor performance and the limited availability of radio channels to support them. They were absolutely thrilled to be able to use a phone that wouldn't be wired to their vehicle.

FCC commissioner Charlotte Reid had been a nationally famous singer and a congresswoman. She had overcome prejudices to serve nearly five terms in the House before her FCC appointment by Richard Nixon. She was, at that point, only the second woman appointed to be an FCC commissioner—and she marveled at the DynaTAC.

Fifty years ago, in 1973, there was no internet, no World Wide Web, no personal computer, no digital camera. There weren't even cordless phones, which were patented only a few years later. Yet we had created the DynaTAC.

AT&T still saw only vehicle-based phones. In a published article in 1970, Dick Frenkiel of Bell Labs wrote that they "assumed an ultimate user group of several percent of the presently registered motor vehicles."[11] In AT&T's view, the cellular phone was seen as merely an adjunct, an accessory to the wired telephone network that would continue to be the centerpiece of communications.

Motorola's advance with the DynaTAC also represented a major step in the development of cellular systems. What we did was much more than a handheld phone—we created a system. We were able to evolve Bell Labs' original idea of a cellular system because we viewed cellular through the lens of portability. Based on the CPD work and our experience with pagers, we understood the additional challenges that needed to be solved.

What would happen when a person took their portable phone up an elevator in a building? In this scenario, contemporary technology expected that a portable phone signal would "capture" signals from other phones and interfere with them. If someone made a call from within a tall building, their signal would jump across cells and the seven-cell clusters and interfere with another phone call. The signal from a person in a building was radically different from one in a car moving steadily from cell to cell. For true portability, this needed to be solved.

Which we did. The patent that Motorola received for the DynaTAC included a system of power reduction that would minimize interference. Reduction in power allowed many more subscribers to be served in each city and also extended the battery life of the portable phone. Without that system reduction in power, the system that Ring and Young proposed in 1947 at Bell Labs could not have accommodated portable phones. This form of power reduction is built into every one of the over six billion cell phones in operation today.

If the FCC had given the green light to AT&T's monopoly proposal, mobile telephony would have been chained to heavy equipment in cars, trucks, and trains for decades. Monopoly dominance would have been expansive. If the FCC had assigned 30 MHz of spectrum to AT&T, that would have accommodated six hundred voice channels. Before cellular, the government had never allocated more than a dozen or so channels.[12]

This might be my favorite picture from the DynaTAC episode. FCC Commissioner Robert E. Lee, one of Motorola's biggest skeptics, making a call on the DynaTAC.

My New York City sidewalk call, the press conference, the media coverage, our Washington demonstrations—all of it had the effect we wanted. The FCC decided to study cellular more closely and eventually denied AT&T's proposal for a monopoly in cellular service. Instead, they granted small parts of the radio spectrum for future cellular use, asking for system trials as an initial step. The FCC agreed with Motorola and made the market competitive by mandating multiple systems in different areas, not just one offering. More spectrum would be allocated after the trials. The FCC also set aside 20 MHz for Motorola and others for two-way radio service. My favorite moment was seeing FCC commissioner Robert E. Lee, enduringly skeptical of Motorola, making a call on the DynaTAC.

The public was the ultimate winner of the FCC's actions. And the rest, as they say, was history.

Well, not quite.

The implications of what we had done were larger than I initially thought. Our immediate goal, achieved through a harried three months of development, was to show the FCC that competition could better spur innovation than monopoly. But it ended up as more than just a demonstration. It was actually a proposal for an entirely new line of business for Motorola. This was about more than spectrum and more than a new product. It opened the door to an alternate future for Motorola. That was not entirely welcomed by some, and for good reason.

Motorola executives—especially Bob Galvin, Bill Weisz, and John Mitchell—had supported our DynaTAC push and demonstration. There was unanimity within Motorola regarding AT&T: they could not be allowed to extend their monopoly. A competitive market was essential, and we had to prove to the world that you didn't have to have a monopoly to run the cellular business.[13] We had to beat them; we had to beat the monopoly.

Yet others in Motorola were not particularly enthusiastic about the cellular business in general. And, once more, there was considerable disagreement and internal divergence. This time, it was about Motorola getting into the manufacture and buildout of a full-blown cellular system.

For decades, Motorola's biggest commercial success was in two-way radios, and that success was based on selling directly to users. Motorola had one of the largest direct sales forces in the world. Its sales representatives called upon the police superintendents, fire chiefs, plumbers, and electricians who used the radios. The sales representatives, as encouraged by management, knew exactly the unique requirements of each kind of customer. Our corporate culture and success were firmly grounded in these relationships. This was the basis of John Mitchell's reservations about cellular, notwithstanding his support for the DynaTAC sprint. Because of our direct relationship with users and large sales force, John thought we would lose strategic value in the cellular market.

If Motorola were to begin manufacturing cellular equipment, it would be dealing with communications carriers who would treat the equipment as a commodity, with the objective of getting the lowest price. The carrier determines the equipment specifications to ensure compatibility. That puts the carrier in control. This was not a milieu with which Motorola was familiar or comfortable.

John eventually changed his mind. Despite his reservations, he became an enthusiastic supporter of Motorola's creation of the cellular industry. Since John was the one who had inspired me and taught me the value of personal communications, I was particularly buoyed by his support.

Motorola would go on to spend around $100 million in the decade after the first call on everything required to create cellular phones and their infrastructure. We really were betting the company on this demonstration; Bob Galvin saw it that way at the time.

Motorola wasn't alone in its internal strife. Those working on cellular telephony at Bell Labs were also divided as to what would work and how and what direction should be taken.[14] The transition to a new technology, a new system of communications, was bound to create disagreement.

When I spoke to Rudy Krolopp for this book, he told me that, at the time we were doing the DynaTAC work, I was one of the few people who thought that demand for cell phones would grow "exponentially, an infinite amount."[15] Motorola was positioned to lead the equipment market for both cellular handsets and for the radio and switching equipment that made the handsets work. There was a huge amount of effort necessary to get the FCC and the industry to adopt rules and standards that would not constrain the growth of the cellular telephone industry. Without the creativity and persistence of people like Don, Rudy, Roy, and others who turned a vision into reality, the communications world would be very different.

EUREKA DOESN'T HAPPEN

For the nineteen years I worked at Motorola before the development of the DynaTAC and the decade following, I was continuously challenged in an open-minded environment and with an extraordinary degree of freedom.

For decades, "[The Galvins] created an environment that drove people to invent and fail and learn and invent again. Motorola became known for its culture of risk taking, its investment in training and development, and its almost fanatical insistence on respectful dealings among employees."[1]

During his three decades as Motorola CEO, Bob Galvin emphasized creativity and idea generation. Motorola, he declared, "whenever and wherever possible . . . tried to go the other way."[2] Bob encouraged development of processes that would make invention and innovation more likely. That is not a contradiction. Innovation is a process that can—must—be managed.

Motorola encouraged inventors with nominal cash rewards for patents applied for and granted, public recognition, and a unique promotion path for inventors. The recognition was most important to us. We had a patent award dinner every year, at which the engineers were treated as heroes. But the real reward was seeing one's ingenuity converted into products that generated revenues and made people's lives better. Motorola's encouragement paid off in the form of product leadership but also in a portfolio of thousands of US and international patents.[3]

It is unlikely, perhaps impossible, that there has ever been a truly original invention. It is also unlikely that we can ever really know who "invented" something in the sense of being the first to envision or assemble a device or process. Alexander Graham Bell is generally credited as being the inventor of the telephone. But he was in a patent filing race with Elisha Gray. In Italy, Antonio Meucci had already filed the design for a "talking telegraph." They were all indebted, in turn, to the existence of the telegraph and work done on what was called the phonautograph.[4]

Radio, as a facilitator of communications, is more complicated—the credit for inventing radio depends on geographic location. Guglielmo Marconi was educated in Britain and is recognized in those countries, and in many others, as radio's inventor. Marconi led the commercialization of radio and deserves a huge amount of recognition for that. In the United States, thanks to a 1943 decision by the Supreme Court, Nikola Tesla's radio patents are given primacy. In Russia, Radio Day is celebrated on May 7 to honor the invention of radio by Alexander Popov. In India, Jagadish Chandra Bose demonstrated radio in 1894 but had no interest in patenting his invention.

Predating the phone and radio inventors were breakthroughs in electromagnetism by people such as Michael Faraday, André-Marie Ampère, and Heinrich Rudolf Hertz. Invention always builds on previous invention. For the DynaTAC, we integrated many individual technologies that were already in development but hadn't been put together in a single device before. Don Linder later observed, "We had a running start because we were doing our homework [for all of the technologies inherent in the phone and the remainder of the system that made it work]. This doesn't spring out because someone was sitting in their office saying, 'We need a portable phone.'"[5]

That's how invention happens. There is rarely a "Eureka!" moment. It may take a lifetime (even generations) of learning, successes, and failures to develop the judgment and self-confidence needed to make a new idea real. Creation of the handheld cell phone was, technically speaking, inevitable. The cell phone—as a commercial, consumer product—would still have happened without the contributions of me and my colleagues at Motorola. But it would have taken longer, and it would have been quite different. Someone needed to nudge it forward into existence.

Experiences and skills, like inventions, build on each other—and when the moment arrives, they coalesce to become the essence of successful innovation. It's hard, perhaps impossible, to know where experiences may lead. I had no secret sauce in my career, no long-range plan. I immersed myself in every project and did my best. That approach served me well, but it's not for everyone. When I became an avid competitive runner (at around forty-nine years old), I understood the concept of pacing oneself to retain enough energy to finish the race—and I totally ignored it. My strategy was to run as fast as I could the entire time. That strategy worked superbly for 10K races, but I never completed a marathon. I did participate in the Chicago marathon one year. My idea was to get into the excitement and drop out at the two-mile mark as the race passed my house. But by then I was too swept up in the energy to stop. By the time I ran out of steam, I couldn't find a taxi, because so many roads were blocked off. Finally, around mile thirteen, a spectator asked me where the finish line was so he could meet his wife. I offered to direct him to there and give him a good parking spot in my driveway if I could ride along in his car. The finish line was right near my house, too.

I learned at Motorola that it is possible to encourage and inspire others to invent and help create practices and rewards to stimulate the process. It's a lot easier to discourage the spirit of invention. I can't tell you how many people told me that the idea that anyone would want a handheld portable telephone was simply preposterous.

In any innovation project, you're going to be faced with a situation of having to manage and motivate others—some of whom don't report to you. You will be required to get them to accomplish things as a group that none of them individually thinks is even possible. One of the most effective techniques that worked for me was setting a fantastic future in front of people and challenging them to create it. Let's create the first cell phone! This was something I did for myself, as well. This is the essence of being a dreamer.

Another skill in managing innovation is in identifying and acquiring the necessary resources, human and otherwise. Technology has become too complex to allow individuals to complete important developments on their own. Progress requires teams and organization and leadership to make anything happen in the world of technology. It requires building relationships with people everywhere, inside and outside. And it requires building support from management.

This support, in turn, requires that management have a vision of where technology is going (or might go) that will allow them to evaluate innovative projects on the basis of more than just the immediate financial consequences.

⌒⋇⌒

Time and again at Motorola, we were faced with opportunities to sell a new product or win a contract. We never did just the minimum. I always asked myself and others the what-if questions: What if we manufactured a perfect quartz crystal?

What if we combined separate receiver filters? What if we only had three inches of space to work with? And so on.

The basis of these questions must be an understanding of the underlying science, of the needs of society, and the judgment of discriminating between what can be done and what can't. There is nothing like the challenge of creating an alternative world, a new future of possibility. Once that vision exists, it's possible to persuade teams of executors to achieve what they once believed impossible. And if you run into resistance, turn the question around on the skeptics: Why not?

In that same spirit of turning questions around, it's important to remember that innovation is not always about new products, it may also be about new processes, even processes that reclaim, reuse, and repurpose what was once itself a new innovation.

My friend Barry Sweet is a third-generation junk man, which means he's in the sustainability business at a time when society is in a sustainability crisis. Our oceans are loaded with plastic, garbage dumps are becoming scarce, and we haven't figured out how to get people to pay for the hidden costs of their discarded cell phones, computers, and routers. For three generations, Barry's family has bought people's discards, processing and repurposing them—from rags to scrap metals to electronics. Now Barry works for a Finnish company in a Midwestern town. The company disassembles electronic equipment of all types using processes that range from hand labor to huge machines using knives and hammers to crush the equipment, magnets to separate the steel, and chemicals to dissolve and reconstitute the copper. In an amazing reprisal of his grandfather's process, residual amounts of gold and silver that came along with the copper can make the difference between profit and loss. Other processes recover cobalt, nickel, and molybdenum as used in creating specialized forms of stainless steel.

Innovation has always been about standing on others' shoulders and about collaboration. Innovation is not the province of a select few. It's open to anyone. Yet accepting this notion is part of our challenge. We too often treat innovation as an elite activity, and, for everyone else, we have an "expectation level that is too little compared to what is contributable."[6] Instead, we ought to encourage wider participation in the innovation process. That's essential if we are to realize our bright technological future.

PUTTING CELL PHONES
IN PEOPLE'S POCKETS

M y call to Joel in 1973 was a landmark, but it was still many years before we could put a cell phone in everyone's pocket.

After 1973, the war shifted to a technical battle as Motorola fought to inject its superior technological understanding of radio into the specifications of the new cellular industry. Our original projection, that "installation of the first system will be completed in New York City by 1976," proved to be optimistic.[1] The FCC spent a very long time drafting rules and regulations, which contributed to the delayed arrival of cellular systems. Creating the technology to make portable telephones work was also a major challenge and took much longer than anticipated, contributing to the delays as well. Successive versions of the portable telephone were smaller, lighter, and more manufacturable—but this took time. Over the next decade, Motorola's research team built at least four versions of the DynaTAC. The biggest technical issue was that we couldn't produce a commercial product with a thousand parts in it that required an engineer to be standing next to you each time you made a call. We needed something reproducible.

After the excitement and thrill of 1973, enthusiasm inside Motorola waned within a few years. By the middle of the decade, we were spending millions of dollars on cellular system development and had gone through a couple of generations of it—and still had no revenue to show for it. My group within the

Communications Division continued to invest in and develop new versions of the DynaTAC handset, fixed radio stations and switches, funded largely by its profitable IMTS business.

There were also broader cultural barriers to acceptance of the concept of a radiotelephone replacing the wired phone. As one commentator later framed it, "Why would anyone pay a monthly subscription fee and hefty per-call charges when 10-cents-a-call phone booths were everywhere?"[2] AT&T had actually hired a market research firm to figure out if people would be willing to pay for portable telephone service. They reported no interest and little conceivable market.[3] AT&T continued down its car phone path, anxious to prove its contention that only AT&T had the technology and money to create a system of vehicle-based phones.

By 1979, both Motorola and AT&T were testing cellular systems: AT&T in Chicago using car phones, and Motorola in Baltimore and Washington using handheld portables. Some of the car phones used by AT&T in their trial were built by Motorola. But Motorola had no significant revenue from cellular.

❧

In 1980, John Mitchell became president of Motorola, and in 1981, the FCC announced the allocation of frequency in the 800 MHz range for cellular telephony.

In 1982, Motorola CEO Bob Galvin visited the White House, along with other business leaders, to discuss trade policy with President Ronald Reagan. He arrived early and dropped in on Vice President George H. W. Bush. They were old friends, going back to when Bush headed the Central Intelligence Agency and Bob was an advisor to that agency.

Galvin brought his grandson as well as a working version of Motorola's DynaTAC cell phone. At that time, we were a few years into our FCC-authorized trial of a cellular system in Washington and Baltimore. Motorola's system had hundreds of handheld portable phones deployed, as well as many car phones. AT&T's trial in Chicago, Motorola's home turf, deployed only car phones.

Vice President Bush didn't recognize the "brick" phone that Galvin was holding. Galvin suggested, "Why don't you call Barbara?" Bush called his wife at home and said to her, "Guess what I'm doing? Talking on a portable telephone."

The vice president handed the phone back to Galvin and asked, "Has Ron seen this?" Meaning President Reagan. After the trade policy meeting later that day, Bush urged Galvin to linger and said to the president, "Ron, you've got to see this thing."

The president looked at the brick phone in Galvin's hands and asked what he was carrying.

"Mr. President, this is a portable phone. It should have been on the market. It will be out pretty soon," Bob told him.

Reagan took the phone and made a call.[4] He asked Galvin, "What is the status of this thing?"

Bob was in general a modest man—but he was also a chief executive, ambitious, driven, and not about to let an opportunity like this pass.

"Well, Mr. President, it's languishing. We are ready to take this to market now, but the FCC hasn't figured out how to select the operators who will provide the service. The phone company wants a monopoly; we're hoping the FCC decides that competition would be a good thing. Maybe they're holding back because they want the Japanese to be the first ones to enter the market."

Reagan winced at Galvin's obvious manipulation but laughed and turned to an aide: "You get ahold of the FCC chairman and tell him I want this thing released."[5]

Galvin had a lot at stake in delivering his message to the White House. He had led Motorola to invest over $100 million during the previous decade without a penny of revenue. He bet much of Motorola's future on the success of the product. Within Motorola, there were naysayers aplenty. One could hardly blame the businesspeople who worked hard every day to maintain Motorola's dominance and profitability in the two-way radio industry for questioning the viability of this new and untested product.

Much of that investment was in the technology I've mentioned already, as well as lobbying the FCC and Congress. Development of the portable phone that Don Linder, Rudy Krolopp, and many other Motorola engineers pioneered was an enormously costly task. The breakthroughs that we demonstrated in 1973 had to be made manufacturable at a cost low enough to be marketable. Between the 1973 introduction and the Washington trial, we had created at least four iterations of the brick, each more robust and more manufacturable.

The DynaTAC was useless without the switching equipment and radio base stations that made the portable work. I established a new department in my systems operations group led by Andrew Daskalakis, whom I had hired away from Bell Labs. Andy was a first-class engineering leader, who proved to have management skills comparable to the talent he had demonstrated at the Labs. Andy flourished at this entrepreneurial challenge. He brought in others from the Labs, like Phil Porter, also a superb engineer.

By the time Bob Galvin visited Bush and Reagan, Motorola had finally achieved a manufacturable portable phone and the cell site equipment to make it work. We were ready to go—and so, at last, was the FCC—and we wanted to make sure the FCC continued on a path to competitive cellular service and portable telephony.

It had been an arduous regulatory slog for the commission and everyone involved. The FCC had asked AT&T for a cellular spectrum proposal in the late 1960s. When Motorola introduced the DynaTAC in 1973, we had told the world we'd have cellular service up and running in just a few years. Our optimistic pledge was based on expectations of breakthroughs in phone technology and swift FCC action. That was naive. It took four years for the FCC to authorize experimental systems by Motorola and AT&T in one city each. We ended up needing most of that time to refine the engineering design of the DynaTAC portable and the infrastructure equipment that made it work.

Within months of Bob's visit to the White House, the FCC began authorizing commercial licenses for cellular service. In late 1983, Motorola and AT&T began serving commercial customers. Bell was first in Chicago in September. A picture from the AT&T archives shows "the first cellular customer" making a call—on a car phone.[6] The DynaTAC portable was also approved by the FCC in September of that year, enabling Motorola to get started in Washington in December. Motorola's first commercial cellular system operated in Baltimore and Washington beginning in November 1983. Ten years after my New York City sidewalk call, the first commercial cellular phones went on sale for nearly $4,000. That would be like buying a phone today for over $10,000.

Within three years, two competing cellular systems were operating in America's ninety largest markets.[7]

Why did it take so long? Some analysts believe that the FCC deliberately decided to "slow roll" the introduction of cellular service by limiting how much radio spectrum was available for it.[8] It was slow, but I don't believe this is the full story. The FCC faced a series of hard problems under severe and divergent pressure from powerful lobbying groups including AT&T, television broadcasters, radio common carriers, and, of course, Motorola and its allies. Other countries—Japan, for example—rushed their technical and regulatory decisions to facilitate an early start. In the end, the early starters had to revamp both their technology and regulatory approaches to accommodate portable phones, to adopt US technical standards, and make their cellular service competitive. The Bell System was ready to go with car telephones several years before the handheld phone technology was ready. The FCC's delay ultimately resulted in lower prices, better service, and the simultaneous introduction of handheld and car phones.[9]

The difficult regulatory rollout of cellular telephony was complicated by another momentous event. After all its lobbying and technical efforts, it was not AT&T who finally offered commercial cellular service alongside Motorola. In fact, it was one of the "Baby Bells"—ironically, Illinois Bell, the Bell Operating Company in Motorola's backyard of Chicago. The federal government had prevailed in antitrust action against the Bell System. That lawsuit was originally brought by the Department of Justice in 1974. Nothing moved quickly in Washington when it came to communications policy.

Even as AT&T was on the brink of triumph—securing a hoped-for cellular spectrum monopoly—it faced mortal peril. It was the largest company in the world, employing one million people and epitomizing successful corporate vertical integration. It controlled local telephone service, long-distance service, and the manufacture of telephone system equipment, all supported by a highly regarded research arm. None of this proved strong enough, and by the end, the Bell System acquiesced to its own breakup. A federal court ordered divestiture to become real at the beginning of 1984.[10]

After a decade of litigation and regulatory delays, the landscape of American communications was transformed almost overnight. Within the space of just a few months, handheld, portable cell phones were available commercially, and one of the most dominant companies of the twentieth century—and one of the biggest and most persistent presences in most Americans' lives—was in pieces. The cell phone business was relegated to those pieces, willingly abandoned by AT&T as a "throwaway."[11] A former executive said AT&T was "more interested in the Yellow Pages at the time than wireless."[12] Such was the extent of monopoly-based thinking, informed by high-priced consultants. Ten years later, to rejoin the cellular market, AT&T had to shell out $12.6 billion to acquire McCaw Cellular Communications.

⌒⋇⌒

Joel Engel disagrees that my DynaTAC call to him in 1973 made any difference.

> It didn't register as a big deal. It didn't affect the FCC. At the time, people thought handheld was a novelty. People thought the main use of cell phones was in vehicles. None of us—the FCC, Motorola, AT&T, anybody at that time in the 70s, did not anticipate these things. We thought the business was going to be purely business usage—real estate agents, home repair, people who were in their vehicle a lot. We didn't anticipate teenage kids using cellular phones. We didn't anticipate personal residential use. We also didn't anticipate they'd be handheld pocket-sized units. No one predicted the advances in battery technology that's really made it all possible. Handheld was thought a small niche product. We completely missed the individual usage.[13]

Well, of course they didn't anticipate pocket-size units and teenage obsession—few did. I knew that the phone would get tiny; that had happened in pagers and even watches. But teens being obsessed, or even having phones? I might have guessed it for a generation or two later (and it did take a generation for that to happen).

Not many people in 1973 could have predicted use of a cell phone to map your location or make a restaurant reservation. Technology always evolves in directions that its inventors can't (or won't) foresee because of path dependence, blinders, arrogance, or other related reasons.

Focusing on all the things we did or didn't anticipate misses the point. Innovation is much more than technical achievement. It requires recognizing and understanding inherent characteristics of users—and making technology fit those characteristics, rather than work against them. Customers and users will then be your partner in taking technology forward into uncharted territory.

The DynaTAC demonstration benchmarked the new technological frontier. Many of the things we did to put the phone together and make it work—receiving and transmitting at the same time, four hundred channels, power reduction—were industry firsts. Motorola's experience with pagers, portable handsets for police and fire departments, and building equipment for car phones had shaped our philosophy. The company's entire outlook was grounded in making things more and more portable for forty years at that point.

Our patented system for power reduction in the portable phone meant that it could be used in tall buildings or crowded areas. This densification enabled further generations of ever-smaller phones. Power control not only made portables work but also increased capacity, ensuring that the Law of Spectrum Capacity would continue. With this, we changed the entire idea of personal communications. People call people, not places, which means their phones need to serve them, not the other way around.

People are naturally mobile—that was Motorola's belief in the 1960s and 1970s and has been mine ever since. For one hundred years, people wanting to talk on the phone were constrained by being tied to their desks or their homes with a wire. Then they were trapped in their cars. That was not good enough. People want to talk to other people, not cars or offices, homes or phone booths.[14]

My own evolution at Motorola included work spanning these developments in portable phone technology and more.

As vice president and director of systems operations for the Communications Division, I got to start the cellular business, participate in the Motorola challenge to AT&T at the FCC, and stimulate the creation of the trunked mobile radio system that was becoming the heart of Motorola's communications business. It became evident to me, and to management, that vision was my strength. I was good at starting businesses and being involved in the technology, but I had limited interest in or talent for the details of financial management. In keeping with Motorola's inherent objectivity and respect for technology, they created a new job specifically for me, just as they had in making me product manager of portable products, then head of systems operations.

So, in 1978, I was promoted to the newly created position of vice president and corporate director of research and development. Among other things, I oversaw work in the Motorola Integrated Circuits Research Laboratory (MICARL, which came from the semiconductor group), and once again, the quartz crystal manufacturing business reported to me. MICARL and my corporate R&D team provided state-of-the-art integrated circuits to the DynaTAC development program. The corporate team, still being managed by Roy Richardson and with Don Linder's leadership, continued to build successively more advanced portable telephones. MICARL also created the first automobile engine management large-scale integrated circuit for Ford. And I inherited and supported Motorola's AM stereo effort, which became a market leader for a time.

My new corner office (identical to the one created by the Plantation, Florida, architects) on the top floor of Motorola's corporate headquarters in Schaumburg, just outside Chicago, should have been the pinnacle of my career. But to me, it was a pyrrhic victory. I was now a member of the corporate staff, with no authority. I knew, from experience, what the people in the operating divisions thought about corporate staff. When I ran the systems division and the corporate guys made suggestions, the standard answer was "Thank you! That's a great idea; we'll look into it immediately." Which translates as, "Get lost. What do you know about running a business?"

⌒✳⌒

Well, I knew a bit about strategy and learned more in my new position.

Strategy is a term of warfare. The concept of strategy applied to business has companies pitted against each other in a continuing battle. The winner combines profitability, market share, return on assets, and other criteria of business success.

Strategy is what distinguishes an organization from others. It is not planning, wishing, hoping, or desiring. In a well-focused company, strategy lurks in the background, influencing every decision by every manager and employee. A company without a clear, articulated strategy is adrift—it may do well in the short term, but success becomes a chance event. Strategy must be embraced and owned

Me with Motorola's long-time visionary leader, Bob Galvin, June 1968.

by everyone in the company. The most important components of Motorola's communications business were deep understanding of customer needs, direct engagement with customers, paranoia about competitors, and objectivity about the strengths and weaknesses of the company.

After I was promoted to corporate, we began the practice of articulating the strategies of our various businesses in a vehicle called the Technology Roadmap. Bob Galvin originally envisioned the concept, which was intended to stimulate adoption of new technology at a faster rate than our competitors. Bob's idea was that "technology is advancing, and we should have some idea of where it's going and how we should respond."

Management of the process of creating Technology Roadmaps fell to me in my role of VP and director of R&D for Motorola. Bob had his long-range vision. John and Bill had their own versions. I was in the middle, executing implementation by all the operating divisions. With the unswerving support of all three, we created a system that was an ingredient in Motorola's technology leadership for many years. The Technology Roadmap became an integral part of the company's long-term planning process.

At Motorola, thanks to excellent mentors, I had the chance to manage entire projects from start to finish. From a career perspective, no matter what kind of work you're in, this is incredibly important. Going through a project gives you an understanding of the entire process—which you'll need if you're going to be in any sort of management or leadership role. But that kind of opportunity doesn't just fall into your lap. You have to seek it out. I was able to use my experience in creating and conducting the Technology Roadmap process.

The Roadmap became a compilation of predictive tools, which operating units were required to apply, maintain, and report to top management routinely. Bob, Bill, and John reviewed every Technology Roadmap in the company in face-to-face meetings with the units.

Getting to those face-to-face meetings was always a thrill. Bob, Bill, and John would each fly in their own jets (for corporate disaster planning purposes). We'd meet at the airport, and their pilots would use their flying skills to race to be first at the destination. Over the course of the Roadmap review process, we'd fly to Arizona, Florida, and Texas.

I felt like I was in the stratosphere, arriving by limo at the bottom of the stairs leading up to the plane's door. I most often went in Bob Galvin's personal plane. He liked me to fly with him, because then he could charge back the company for the price of a business-class ticket. For all his money, he watched finances closely. Every time I was on his plane, he'd remind me not to drop peanuts on the floor. (I'm not the neatest person in the world.)

Even twenty years after I'd left Motorola, there were times I was invited to heady dinners at Bob's house in Vail, involving curated discussions with luminaries, and he would regress to treating me like his junior engineer, snapping at me, "Pay attention, Marty," if he thought my contribution wasn't on point.

The Roadmaps were not perfect in their predictions, but they made people think. Where they were taken seriously, they were remarkably successful. Bob believed that the Roadmap offered an "amazing ability to see what was coming down the pike."[15]

The Technology Roadmap didn't replace vision as a means for predicting the technological future. And it didn't always prevent miscalculations. Dan Noble was a visionary executive at Motorola, having founded the company's three most successful divisions: communications, semiconductor, and government products. He was respected by management and revered by the engineers. By the 1970s, he no longer had an operating role, but he had not given up thinking about future technologies. Based on that, Dan envisioned innovation in a consumer product with which Motorola had little experience: wristwatches. He was convinced that

Dan Noble, vice chairman of Motorola in the mid-1960s.

a quartz-driven electronic watch would ultimately replace the spring-driven mechanical watch. And I believed him.

On one occasion I sat in the company cafeteria with a group of colleagues and asked, "Anybody know what time it is?" The response, as was typical in those days, was a range of times with a spread of at least five or six minutes. "One day," I predicted, "I will be able to ask that question, and everyone will be within a few seconds of the correct time."

The heartbeat that would make these electronic watches accurate would be a quartz crystal resonator. Believing that Motorola should diversify into becoming a watch manufacturer, Dan was allowed to start a new division, for which he hired engineers and marketers and developed a family of quartz watches.

The crystals for use in a watch were very different than those used to tune mobile radios. They had to be much smaller and operate at much lower frequencies since the objective now was to accurately count out seconds. The watch crystal was essentially a tiny tuning fork; some of the early versions actually looked like tuning forks. A crystal would vibrate at a precise frequency and electronic circuitry would modify the frequency from hundreds of kilohertz to seconds, not unlike what the gears in a mechanical watch do. We managed to design a crystal that, combined with a temperature compensating circuit, kept a watch accurate. With that, we launched a watch crystal business.

The timing was ideal. The superior manufacturing process that my team developed in the early 1960s was still first in its class. Bob Nunamaker was running this business. Although he was an engineer, Bob was a skilled business manager, and he could execute. He acquired the know-how and people. Traditional watch companies created quartz watch product lines, and the nascent computer industry took an interest in the market. We made watch crystals by the millions and were the largest manufacturer in the world with 65 percent of the market.[16] We were selling crystals for two dollars apiece and making a bundle. And I was now in a glamorous consumer products business. I suggested, during a meeting with the president of Timex, that watches no longer had to be circular. Within a month, Timex had a rectangular quartz watch in their product line. Charlie Sporck,[17] CEO of National Semiconductor, met with me to beg for a higher quantity of quartz crystal for his emerging (and ultimately failed) watch company.

Reality set in quickly. Japanese suppliers began undercutting our prices—and not by a small amount. Within a year a watch crystal was selling for fifty cents, well below our cost. We moved our manufacturing facility to Tijuana, Mexico, to reduce costs. John Mitchell realized that while the Communications Division was good at making custom crystals for two-way radios, high-volume manufacturing was different. He transferred the quartz crystal business to Motorola's Semiconductor Division.

Strategic planning for an organization requires clarity of mission, goals, and objectives. But the most important thing is an understanding of its strengths and weaknesses. What I've found is that, too often, people and companies misconstrue their weaknesses (such as the "I work too hard" or "I'm too nice" that are staples of job interviews) and overdo their strengths. It's not enough, for example, to simply believe you're better than others. There has to be some substance to it; that's what the Technology Roadmap did. I lost sight of these guidelines in my failure with quartz crystal watches.

You can be certain that specific predictions made in a road map or plan will be wrong; that's true of any business plan. But having the plan helps organize resources and prepare for change—or failure.

⌒✦⌒

As good as the Technology Roadmap process was (and is when I use it today in advising companies), it was not enough to stem the internal conflict that increasingly divided the company. In the early 1980s, Motorola was at the pinnacle of corporate success—we had at last pioneered commercial cellular service and a portable handheld phone.

By the 2000s, just as John Mitchell had foreseen, Motorola's cellular mojo was gone.

The cellular business lost its way. The story of how that happened is as fascinating as the creation of the industry; perhaps I'll get around to telling my version of that story in another book. In brief, Motorola had been the dominant manufacturer of cell phones and infrastructure equipment in the 1980s. In a catastrophic and meteoric collapse, Nokia and Ericsson ate Motorola's lunch in cell phones and infrastructure, respectively. The company's valuation rose steadily through the 1980s and 1990s, collapsed just before the dot-com bust, recovered a bit, then plummeted again.

On January 4, 2011, Motorola Inc. separated into two independent, publicly traded companies: Motorola Solutions Inc. and Motorola Mobility Inc. Motorola Solutions (NYSE: MSI) provided mission-critical communication products and services for enterprise and government customers. Motorola Mobility (formerly NYSE: MMI) made mobile cellular devices and cable video management equipment. In May 2012, Google finalized its acquisition of Motorola Mobility for $12.5 billion. Not quite two years later, Google sold the residue of Motorola Mobility to Lenovo for $2.9 billion.

When other details were accounted for, Google had, effectively, paid about $3.5 billion for Motorola's patent portfolio. When I had visited the Motorola plant in Libertyville, Illinois, in the mid-1980s, there was a display of hundreds of Motorola's patents. Separated from the rest, in a place of honor, was the patent for DynaTAC. By the time Google bought the patent portfolio, our patent protection had expired; its value was zero.

Today, Motorola Solutions is a multibillion-dollar company, a leader in the field of land mobile and two-way radio communications. In other words, the core business of present-day Motorola Solutions was the core of what made Motorola so successful in the 1950s and 1960s.

There were many reasons for Motorola's decline, and I'm only aware of a few of them. A big factor was hubris—arrogance drove the managers of the cellular business to abandon the diligently honed strategy that had generated the communications business's success. The company's leaders drifted away from its strategic imperative of dealing directly with customers and of knowing customers' business better than they did. Success in two-way radio and the development of the cellular phone fostered arrogance among some of the company's new leaders. They thought they could control the market rather than following closely with customer needs. For example, when the digital revolution began to overtake cellular, some Motorola executives clung to analog technology. This may have been some executives' interpretation of Bob Galvin's dictum that "staying put, at times, is in fact a further manifestation of going a unique way."[18] The marketplace had already decided that digital cellular telephony was the way to go, and customers, the cellular operators, were not interested in the technical nuances of Motorola's position that there was still life in analog cellular phone technology.

Motorola Solutions, on the other hand, nurtured and adjusted the strategy that was responsible for the communications business's leadership. That's part of the reason it is still today a successful and profitable corporation.

FULFILLING
MY FAMILY LEGACY

In the late 1970s and early 1980s, I was terribly frustrated, but I labored on for several years. I started to think seriously about leaving Motorola. It took me some time to figure out my next steps. I was struggling through a period of big changes in which I lost my first marriage. This was also the period during which my horse, Cash, died. I was feeling, too, the tug of family history: a multigenerational legacy of entrepreneurs to the core, a standard I had yet to achieve. After all, I didn't want to be like Harry, my aunt Rose's husband and the only man on my mother's side of the family to have worked for someone else.

I finally left Motorola in 1983, just as the cellular system licenses were being awarded by the FCC and the first commercial service embracing handheld cell phones went live. I moved out of my palatial office in Schaumburg to a tiny cubbyhole in an ancient building on Chicago's lakefront. My one window faced out onto the fire escape. After thirty years as a corporate employee, I shifted gears and became an entrepreneur. With various partners, I helped start a series of businesses in the cellular industry.

As I was thinking about leaving Motorola, I was approached by Russ Shields, a serial investor in technology startups. He proposed forming a company to take on billing and information systems for the new cellular industry, specifically the non-wireline companies, who were not already served by large wired telephone companies. We first offered the opportunity to Motorola. But I made it clear that I was going to work with the new company, whatever Motorola's decision was. After an exhaustive analysis, Bob Galvin said, "Marty, I'm going to let you go do this on your own, because we'll screw it up." The next thing I knew, I was being toasted at my goodbye party.

Founded in 1983 with Russ Shields and Arlene Harris (my second wife and a technology innovator), Cellular Business Systems developed billing software for carriers. In 1986, by which time the company had won 70 percent total market share (including wireline companies), we were forced by circumstances to sell the company to our competitor Cincinnati Bell. Our largest investor, First National Bank, had decided to get out of the venture capital business. Russ was furious and held up the closing until midnight. He had hoped we could hang on to the company until it made us enough for him to buy a baseball club. I, on the other hand, was thrilled. After we'd made some unfortunate hiring decisions, I had ended up as CEO, and, oh boy, I hated the position. That same year, Arlene and I cofounded Dyna LLC, which still exists today and through which we have brought a number of innovations to the market. Except for ArrayComm, the businesses Arlene and I engaged in were all her ideas, each successively more ambitious.

For example, in 1988, Subscriber Computing Inc. developed a system to provide what's now known as prepaid cellular service to low- or no-credit consumers. This was an innovation that ultimately generated an enormous uptake in cellular adoption in emerging markets and developing countries and is now one of the most common ways in which billions of cell phone users access this essential service.

GreatCall, another example, was born out of potential disaster. When a supposedly reputable major company reneged on their contract to manufacture an emergency cellular telephone for one of our companies, SOS Wireless, the company was bankrupted, but Arlene picked up the pieces and formed GreatCall Inc. When GreatCall sold, Arlene gifted some of the proceeds of that deal to investors who would otherwise have lost money in SOS Wireless. Now owned by Best Buy, GreatCall's technology connected the underserved senior market to health services. The company also developed the Jitterbug phone, which prioritized simplicity and ease of use and was included on the *New York Times*' top ten list of greatest technology ideas in 2006.

Separate from Dyna LLC, I cofounded ArrayComm Inc. to develop smart antennae. Smart antennae are so finely tuned and targeted they can give an individual their own infinitesimal slice of the spectrum available. The company has more than 450 patents on a range of technology that is still not fully deployed today. We were too far ahead of our time. In one of those twists of fate, the company is now owned by Russ Shields. During the 2011 Fukushima tsunami disaster, ArrayComm's technology supported the only communication available to many in the area.

There have been other companies and other innovations and successes—and brushes with financial disaster. My fascination and devotion to the world of technology is steadfast. As is my belief and commitment in the social value of these technologies.

I've given up the public company board seats that I enjoyed, because that involved unwelcome travel. I continue to advise several startups and support an R&D project at the University of California San Diego, which I'll tell you about in a later chapter. I serve on the FCC Technological Advisory Committee, for which I spend several hours a week in a continuing, albeit daunting, effort to get the government to stimulate more efficient use of the radio frequency spectrum. In continuing gratitude to my alma mater, IIT, I serve as a Life Trustee.

My passion now, aside from availing myself of the rich body of knowledge on UCSD's huge campus, minutes away from my home, is trying to inspire the youngsters whom we are counting on to fix this messed-up world we live in. The word has gotten around that Marty is available to speak to any group of kids, from the ages of eight to thirty, on a moment's notice. The kids seem to like it; I am thrilled when I sense a spark of curiosity, a desire for learning and an urge to dream that I know can be turned into solutions.

And I dream, too, because that's what I've been doing all my life. That's where Part Two of this book will go: into the future, where technology's capacity to change the world for the better just keeps expanding.

BRIDGE

Marty's Maxims

A s a framework for thinking about the future, I want to offer these principles, which I have developed over my years of working on and dreaming about wireless technology. They guide my curiosity. You've already encountered some of these and will see them at work as we dive into the future in the next part of the book.

- The best way to think outside the box is to not create the box in the first place.
- People connect with people, not places.
- People are inherently, naturally, and fundamentally mobile.
- What we all call "technology" is—and should always be—the application of science to create products and services that improve the lives of people.
- Customization is the inexorable direction of products and services. Every human is unique, different from every other human.
- The radio frequency spectrum is public property in the United States.
- Ubiquitous and affordable wireless connectivity is essential, especially in education and health care.
- There is an abundance, not a shortage, of spectrum.

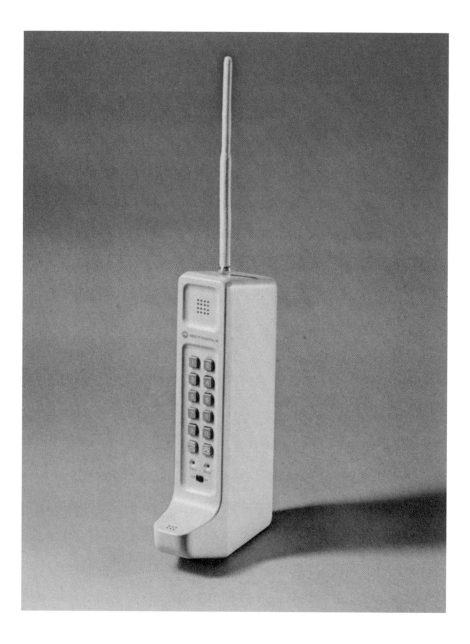

PART II

HOW THE CELL PHONE CHANGES LIVES

I n 2001, roughly 45 percent of the US population had a cell phone. Ownership had doubled in the previous four years and quadrupled over the prior six. On September 11 of that year, terrorists hijacked airplanes and launched attacks in New York, Washington, and Pennsylvania. On at least one of the hijacked planes, passengers used cell phones to communicate with family members on the ground. In many locations, however, cell sites had not yet been installed or existing sites didn't have the capacity to carry the sudden increase in cellular telephone calls. Many first responders and government officials could not be reached, even on the wired network.

On that awful day, radio pagers—what many called beepers—were a principal means of how information about the attacks spread. Even though there were three times as many cell phones as pagers, pagers were still widely used for contacting and alerting people, including at the highest levels of the US government.

Among White House staff traveling with President George W. Bush, "everyone's pager started going off" as word spread of the attacks. There were no phones on Air Force One, which carried the president around the country as they tried to figure out what action to take.[1] The White House press secretary had a two-way pager, not a cell phone, that could send and receive only a few predetermined responses. The presidential entourage was only able to get updates on the attacks by picking up local television signals as the plane flew around. In the North Tower of the World Trade Center, pagers were the principal source of news for those trying to get out. Long lines formed at pay phones around Manhattan.[2]

These pagers were descendants of the first nationwide devices Motorola introduced thirty years earlier. People want and need to be in touch with each other—conveniently, affordably, often immediately, and, during emergencies, urgently. In the late 1960s, when pagers were teaching us about constant connectivity and the portable cell phone was still a distant dream, I had a science fiction prediction. I told anyone who would listen that, someday, every person would be issued a phone number at birth. If someone called and you didn't answer, that would mean you had died. On September 11, we experienced the dark obverse of this prediction—if you tried to get in touch with someone and couldn't get through, you feared they had died.

I expected, even in the early 1970s, that everyone—*everyone*—would want and need a cell phone. Others at Motorola shared this expectation of ubiquity because our two-way radio business had shown us firsthand how many businesses functioned magnitudes better when people were connected. The Mount Sinai providers, airport workers, and Chicago police officers taught us how being connected made organizations work. We remembered the physicians who refused to give up their pagers so we could fix them. Portable devices like the pager and cell phone, through both mundane use and tragedies like September 11, became anytime, anywhere companions, integral to identity itself.

These experiences demonstrated a principle of technology that has shaped my outlook for decades. Proof of a product's usefulness comes when users become so dependent upon and attached to it that they will not give it up, regardless of defects or negative impact. The cell phone has proved this many times over. In a 2014 Supreme Court decision, Chief Justice John Roberts wrote that cell phones "are now such a pervasive and insistent part of daily life that the proverbial visitor from Mars might conclude they were an important feature of human anatomy."[3]

What surprised even me was the speed and scope of adoption. I did not imagine that more people in the world would eventually have access to cell phones than flush toilets.

We tend to overestimate technology's impact in the short run but underestimate its long-term impact. This is known as Amara's Law, after Roy Amara, a Stanford scientist who ran the Institute for the Future think tank for twenty years. Cell phones are a classic example. In Motorola's fact sheet about the DynaTAC produced for the media in April 1973, we said that "the portable phone is designed for use 'on the go,' when one is away from the office or home, where conventional telephones are not available."[4] We believed that most people were "on the go" most of the time.[5] And that is even more true now than it was then.

After cellular phones became a functioning business, the spark that my team and I ignited didn't light much of a fire within the financial community at Motorola. When we prepared the budget for cell phone development, Jim Caile, my marketing manager, showed me a forecast for sales of portable cell phones. We agreed that the first phones would go to market by the mid- to late 1970s. The predicted quantities of product shipments, however, struck me as totally unacceptable.

I knew what it would cost for the engineering and other talents needed to develop a manufacturable cell phone. I had done it enough times, and underestimated those costs enough times, to be pretty confident in my estimates. And I also knew that we would never get our leaders to buy in to a plan that would sell too few cell phones to recover that investment. On the other hand, the naysayers, especially the financial managers, would laugh us out of the room if we were as optimistic as we wanted to be.

I looked at the forecast again. "Double all the sales forecasts," I told Caile, "and let's see if we can sell the plan." He dutifully did that, and management approved.

We weren't that far off on the sales forecast, but only because most of the early cell phones were car phones. The portable was too expensive, and there were not enough cell sites to support reliable portable communications. By 1990, portable performance and size became more practical, and sales grew rapidly. By 2000 it was difficult to buy a car phone; the handheld had taken over. By the 2000s, the collapse of wired telephone subscribers had started. People didn't believe me when I predicted, in the 1970s, that the wired phone would, in the distant future, be made obsolete.

Yet none of us at Motorola envisioned features like cameras on phones. After all, there weren't digital cameras in 1973, so it wasn't even on our radar of technological possibility. Throughout the 1960s, Motorola had been a leader in transistors and incorporated them into consumer electronics. This included the DynaTAC, so we had some notion that, to improve performance, cell phones would include more and more transistors. But we certainly didn't imagine that the cell phone would become a smartphone, a computer in its own right. The personal computer was still in development at the time, and the internet was just being conceived.

Almost universally, predictions about the use and popularity of cell phones were comically wrong.

In 1984, *Fortune* magazine predicted there would be one million cell phone users in the United States by 1989. The actual figure was 3.5 million. In 1994, consultants estimated that by 2004, there would be between sixty and ninety million cell phone users globally. Even the generous margin of error they gave themselves was insufficient: the actual number in 2004 was 182 million.[6]

In the early 1980s, AT&T had commissioned McKinsey & Co. to study the future of cell phones. The company wanted a forecast of usage for the year 2000. The consultancy pointed to all the problems plaguing cell phones at the time: short battery life, high cost, patchy coverage, too heavy. All true. Because of these issues, the total *worldwide* cell phone market size would be around nine hundred thousand by 2000, said McKinsey. That didn't strike the consultants or AT&T, rightly, as very large.

So, in 1983, when the US government won its antitrust suit against AT&T, courts ordered the company to divest itself of most of its operating assets. AT&T thought it wise to include cellular service in the divestiture. Regional companies (the Baby Bells) took over local phone service, while AT&T retained its nationwide network of long lines and switches. Bell management also distributed the incipient cellular business to the regional companies. After all, long lines and long distance were their *real* businesses.

By 1999, near the end of McKinsey's forecast period, "900,000 new subscribers join[ed] the world's mobile-phone services every three days."[7]

It's too easy to poke fun at technological forecasts that turn out to be wildly off base. IBM didn't think there would be much of a market for computers, let alone personal computers. Thomas Edison originally thought the phonograph would be used to send "audio letters through the postal system," while Alexander Graham Bell thought the telephone would be used "as a medium for sharing live music."[8] We don't just underestimate technology's long-term impact—we usually miss completely how technology will be used.

We have the benefit today of knowing just how large the impact of the cell phone has been. And it has been massive. It's unlikely that any reader of this book will be surprised (or even impressed) by statistics about that impact, but it's worth running through some of the most remarkable.

Globally, as I write this, over five billion people have mobile phones—that's two-thirds of the world's population.[9] There are more mobile subscriptions (eight billion) in the world than people (7.7 billion).[10] In the United States, 96 percent of the population has a cell phone.[11] That was zero, remember, less than forty years ago. Cell phones were adopted at one of the fastest rates of any technology in American history, rivaling the adoption rates of the refrigerator and color tele-

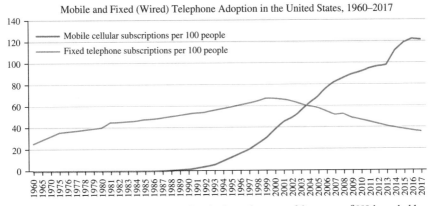

This chart shows subscriptions per one hundred people—over 90 percent of US households had wired telephones. Calculated from data available at Our World in Data, "Technology Adoption." Underlying data from World Bank, World Development Indicators.

vision.[12] One of my favorite charts compares the adoption curves of mobile and fixed telephone subscriptions in the United States.

As we've already seen, mobile phones were primarily car phones in the first few years after the initiation of commercial service in 1983. Handheld phones were costly and, because there was an inadequate number of cell sites, service was relatively poor. By the late 1990s, there were no longer *any* car phones for sale. Meanwhile, fixed telephony has gone into terminal decline and is now at a level last seen in the 1970s.

In a 2012 survey about personal impact, two-thirds of respondents said that cell phones have made it "a lot" easier to stay in touch with the people they care about.[13] That's far and away the largest impact cited. People want to connect.

In 2010, according to surveys about how people spend their time at home, nearly half (43 percent) was spent watching television. One-quarter of at-home time in 2010 was spent on desktop computers, and 8 percent was spent on cell phones.

By 2018, cell phone use at home had caught up with television viewing (33 percent and 34 percent, respectively), and desktop computer time was down to 18 percent.[14] Between 2010 and 2018, overall daily time spent on connected devices (phones, computers, tablets) rose by 2.6 hours—yet daily time spent on cell phones rose by 3.2 hours. Cell phone usage increasingly takes time away from other devices, especially computers.[15] Advertising spending has inevitably followed, with cell phones now accounting for one-third of all ad spending, compared to 0.5 percent in 2010.[16]

Nearly half of Americans (47 percent) say they "mostly" go online with their cell phones.[17] In 2018, digital video viewing accounted for 28 percent of daily video watching, compared to 4 percent in 2010. Most of that digital video viewing is done on phones: in China, six hundred million hours are now spent *each day* watching short-form videos on cell phones.[18]

We still use our phones for their originally intended use: talking to each other. But we also text each other, share photos and videos, watch television, pay bills, check the weather, listen to music and podcasts, make travel reservations, play games, and read books and do research. We no longer need to type our internet queries—we simply speak to our digital assistant. This gives us more time to argue about Android versus iPhone.

The varied uses of our phones stretch the utility of the labels we apply to them. Terms like "cellular" and "cell phone" could only be loved by an engineer—clinical and technical, a name based on reference to the background technological architecture rather than any specific use. Labels used in other countries try to describe function rather than form. In Britain, it's *mobi* or *mobile* (with a hard *i*, of course); in Spain, *móvil* (mobile); in Germany, *handie*; and in Japan, it's *keitai denwa*, which translates literally as "portable telephone." Each of these is more accurate than "cell" phone. Now, of course, we have smartphones and, by implication, not-so-smartphones; for many, simply saying "mobile device" is the best descriptor.

No matter what we call the cell phone today, it still serves the same basic function as it did when we called it a handheld, portable phone in the early 1970s. It enables people to connect with each other. That was the fundamental insight that drove our work at Motorola and continued to motivate me my entire career: connections are made between people, not places. The cell phone offers freedom to communicate with others anywhere, anytime, to and from any place.

This idea—that people connect with people, not places—has become, for me, a core principle of technology and human behavior. One of the most striking illustrations of this principle has been the rise of social media, enabled and accelerated by the cell phone.

Most of us identify social media with the services we use for it today: Facebook, Twitter, Instagram, WhatsApp, WeChat, LinkedIn, and more. It's no accident that these companies—representing the most well-known and successful in social media—were all started in the first decade of the twenty-first century. This was the period known as Web 2.0, when, following the 2001 dot-com bust, new software companies were committed to building a new phase of the internet, with social engagement a big piece of it. Yet it was also a period of accelerated cell phone adoption around the world. Phones were becoming "mobile devices," with

BlackBerry leading the way. That culminated with the introduction of the iPhone in 2007.

Social media was tailor-made for this marriage of software and hardware—applications like WhatsApp, WeChat, and Instagram have been mobile-first, launching straight to cell phones. In a way, this isn't all that new. Social media, according to at least one interpretation, has been around for thousands of years. It's two-way horizontal information exchange among people, passing through social networks and mediated by various forms of technology.[19]

In the nineteenth and twentieth centuries, social media was overwhelmed by hierarchical systems of information and control. Information was delivered rather than exchanged: it was one-way. The cell phone—as a form of people-to-people communication independent of wires and places—helped break open that system. Today's social media is, well, social again, thanks to cell phones.

Social media use is most often associated with young people, but they're far from alone. While 86 percent of millennials report using social media, 59 percent of boomers and 28 percent of the silent generation also report using social media. Those shares among older adults doubled from 2012 to 2019.[20] Parents worry about "screen time" for their children and the amount of time that they spend on their phones, especially teenagers. It is, admittedly, a lot. But what are they doing? According to surveys, 84 percent of American teens are "connecting with other people," and 83 percent are "learning new things." Well over half say they rarely or never use their phone to avoid interacting with people.[21]

Are there negative effects of cell phones? Yes, but that's true of any pervasive technology. I still recall the concern that television aroused in the 1940s and 1950s. If the predictions of the television naysayers had come through, we would be a world of non-thinking zombies with eyes glued to the tube during all waking hours. We survived television. Cellular service is less than forty years old, and the smartphone has been around in volume for less than fifteen years. I have confidence that people are smart enough to figure it out and the net impact of the cell phone will be judged as positive. That is certainly the case today.

TAKING ON POVERTY

South Asia has a labor market challenge. In recent years, hundreds of thousands of new job opportunities have been created, thanks to technology, in sectors such as delivery, retail, logistics, hospitality, and more. Fortunately, there are millions of potential workers available to fill these jobs. But there's a problem.

Many of these workers are migrants: in India, for example, they travel across the country to fill seasonal jobs, sending money back to their families. This means that companies have a hard time managing their workforce—it could turn over entirely within a single year. Recruiting never ends. The other part of the problem is for the workers. There is no good way for them to find the jobs for which they might be good matches. Existing platforms are geared toward white-collar workers, not lower-skill or entry-level workers who fill the new types of jobs.

Enter the cell phone. More precisely, enter an entrepreneur using the cell phone and social media—and demonstrating two of the principles that I've derived in my career—to solve this problem.

In 2016, Madhav Krishna started Vahan, a company focused on improving the job matching between workers and opportunities. Using WhatsApp (and other phone messaging apps), Vahan applies artificial intelligence to gather basic information about a worker, send them job openings, and help them apply. All within a span of mere minutes.[1]

The existence of Vahan was brought to my attention by my friend (and hiking partner) Suren Dutia. A serial entrepreneur himself, Suren also served as CEO of TiE (The Indus Entrepreneur) Global for four years. TiE is an organization that has long helped entrepreneurs from South Asia, such as Madhav Krishna, with networking and support. On our long hikes together along the Southern California coastline, Suren has listened to me (or, perhaps, tolerated me!) talk about my experience with cell phones and wireless communications.

He pointed me to Vahan as an example of two insights that drove our work on portable technology at Motorola and that continued to inform my career afterward. One, which I've already mentioned in this book, is that people are inherently, naturally, and fundamentally mobile. Any product or service that interferes, or tries to work against, that mobility is suboptimal. The second is that what we all call "technology" is essentially the application of science to create products and services that improve the lives of people. That is—and should always be—the goal.

The team behind Vahan is creating technology that works with, not against, people's inherent mobility. Especially in South Asia, with large flows of migrant workers within and between regions, millions of job seekers are constantly on the move. Prior job search tools required them to either show up to a specific place or sit in front of a computer. Companies looking for workers can, with Vahan, harness that mobility. All of this improves lives and, for many of the workers, will help lift them out of poverty. During the global COVID-19 pandemic in 2020, Vahan worked with a large telecommunications company in India to use its services to give workers information on where they could find relief resources.[2]

Vahan is another step in the continued development of the cell phone that has the potential to eliminate, or nearly eliminate, poverty worldwide.

Cell phones have already had a substantial impact on poverty reduction, especially in lower-income countries. Further integration of cell phones into more activities, and continued innovation in that integration, will extend this. The lack of a cell phone is not what has kept billions of people in poverty for so long. But some of the main causes of poverty around the world—lack of access to resources, insufficient infrastructure, absence of (or hard-to-access) jobs, corrupt governments, gender inequities—are directly addressed by the cell phone.

In many lower-income countries, it's been restrictions on mobility—or the presence of institutions that work against mobility—that have contributed to the persistence of poverty. The cell phone doesn't reduce poverty itself, but because it frees people from these restrictions, it contributes to their ability to escape poverty. A wireless connection is still necessary, as is network infrastructure. Organizational changes in the public and private sectors are also often needed to facilitate cell phone adoption and use. People are mobile and want their technology to reflect that; this pushes some of the necessary changes.

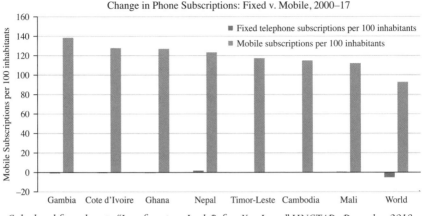

Change in Phone Subscriptions: Fixed v. Mobile, 2000–17

Calculated from data in "Leapfrogging: Look Before You Leap," UNCTAD, December 2018.

Cell phone adoption has been more rapid in emerging markets and developing countries than elsewhere, in part because they started with far fewer wired phones. Low-income countries have leapfrogged fixed telephone lines and gone straight to mobile telephony. The wired network was poorly developed. So, for many people, the cell phone was their first telephone. The above chart, for example, shows that in some countries, growth in fixed telephone subscriptions is almost nonexistent while mobile phone subscriptions have skyrocketed. As in wealthier countries, there are frequently more mobile subscriptions than people.

What's particularly interesting about mobile phone adoption is that—again in contrast to fixed telephone lines—it appears to be independent of income. With fixed telephony, there seems to be an income threshold for countries beyond which fixed telephone subscription rates rise. For mobile phones, however, there isn't any noticeable threshold: even the poorest countries have per capita mobile subscription rates at the same levels as much richer countries.[3]

The adoption curve for low- and middle-income countries is even steeper than in the United States (see next page).

Fixed telephony never achieved the status of life necessity or "feature of the human anatomy" that mobile phones have. In many emerging and developing countries, mobile phones have also passed computers and tablets in how people access the internet.

Thanks to rapid adoption, the mobile ecosystem of technologies and services that has developed around mobile phones generates an estimated $4 trillion in economic value annually.[4] That sounds like a large number, and it is, representing about 5 percent of the global economy. But numbers like those are also pretty

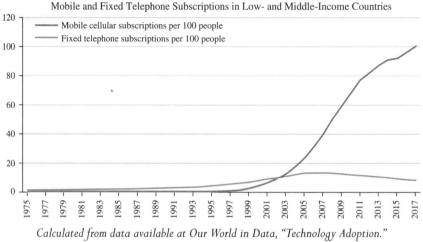

Calculated from data available at Our World in Data, "Technology Adoption."
Underlying data from World Bank, World Development Indicators.

abstract—they don't convey the full sense of how mobile phones have truly transformed lives. Yes, we all spend several hours a day on our phone and use it to me-
diate lots of our daily activities. But it's been in emerging markets and developing
countries—which make up a large percentage of the world's population—where
we've seen the most remarkable impact of the cell phone and the most creative
uses of it.

Today, in many developing countries in Asia and sub-Saharan Africa, cell
phones are filling a key gap in personal identification. Around one billion people
globally have no way to prove their identity—so they have difficulty opening bank
accounts, voting, getting health care, and more.

But most of them have cell phones. And those cell phones have been turned
into a means of establishing and proving identity. Nonprofits, for-profits, aid agencies, and governments are fully integrating cell phones into something as fundamental as personal identity. In Burkina Faso, for example, cell phones are used to
register births; in Uganda, every SIM card in a phone is now registered with an
individual's national identity number.[5] Elsewhere, cell phones are being used for
identity verification in health care and financial services. Most intriguingly, cell
phones have proved vital for identity usage by refugees and the governments and
humanitarian agencies serving them. Many refugees store personal identity information on their phones, and, in at least one instance, the United Nations is using
phone numbers as identifiers to track refugees and better assist them.[6]

Thanks to innovative uses like this, the cell phone has been "the single most
transformative technology" for economic growth and poverty reduction in devel-

oping countries. That's according to Jeffrey Sachs, an economist and author of the book *The End of Poverty*.[7] The cell phone has been one of the biggest contributors to falling poverty across the world.[8] In 1981, 44 percent of the world's population lived in extreme poverty; that was down from 72 percent in 1950. Today, just 10 percent are classified as living in extreme poverty.[9]

This is an extraordinary record and should convince anyone that continued progress is not only achievable but also likely. There are many reasons for it. In the nineteenth century, it was the Industrial Revolution in Europe and the United States, then Japan. In the twentieth century, the biggest economic story was growth in East Asia and Latin America. China alone has accounted for a huge share of global poverty reduction. In each case, the drivers of growth and poverty reduction can be reduced to a fairly simple formulation: "If people have freedom and access to knowledge, technology and capital, there is no reason why they shouldn't be able to produce as much as people anywhere else."[10] Remember my observations above: people are inherently mobile; people connect with people, not places.

The cell phone turbocharged poverty reduction because it symbolizes and provides the freedom of mobility that people want. Given that freedom of mobility, they use it to create, innovate, and improve the lives of others. We've seen this play out in several different ways.

Economic growth. Cell phones boost overall growth. A 10 percent increase in mobile penetration can raise a low-income country's annual GDP growth rate by as much as 1.2 percentage points.[11] A single percentage point may sound like a pittance, but consider the "rule of 72" used by economists and investors. A country with an economy growing at 2 percent per year will see its economy double in size in thirty-six years. Mobile phone penetration raises the annual growth rate to 3 percent: now the economy will double in twenty-four years. That's an enormous impact on people's lives. For individuals and entire communities, mobile phone access has been shown to raise incomes by as much as 20 percent.[12]

Mobile money. There are today nearly one billion registered mobile money accounts globally—half of them in sub-Saharan Africa.[13] In many countries, mobile money—the use of cell phones to transfer funds, make payments, and store savings—far exceeds traditional banking.[14] In India, the Unified Payments Interface (UPI) is a massive digital payments system based on mobile devices and messaging apps. It played a key role during the COVID-19 pandemic in 2020 in helping the Indian government provide assistance to citizens.[15]

Mobile money has helped bring millions into formal financial services, allowing them to build assets and reducing risk from disruptions. Financial inclusion, driven by mobile money, "is one of the most important ways to bring people out of poverty."[16] Using cell phones to transfer money helps migrant workers (both within and between countries) send money—and hope—to their families. Such remittances can be a huge source of income for many in developing countries. Mobile money also helps individuals build up credit histories, which allow them to access additional financial products.[17]

Reducing corruption and improving public services. In many countries, cell phones have been put to use in improving delivery of public services. In Pakistan, simple cell phone calls to school officials about their duties led to an increase in school enrollment.[18] In India, the reverse strategy was used. Farmers were called on their cell phones to see if they had received government payments— and the relevant government officials were told the calls were being placed. This had the effect of increasing the likelihood that farmers received the payments they were due.[19] Not incidentally, these uses of the cell phone have helped reduce corruption among government officials.[20] According to the United States Agency for International Development (USAID), when they worked with the Afghan government to pay public employees through mobile phones, "it cut out so much graft that some employees thought they were actually getting a 30 percent raise."[21]

Access to market information. For fishermen and farmers (who make up a large share of the poor in developing-income populations, especially in rural areas), cell phones allow them to stay in touch with each other and learn about market prices for their goods. This helps them negotiate better prices with wholesalers and distributors. In Honduras, access to market price information in more than one city—via text messages—enabled farmers to receive higher payments than those who did not receive the information.[22] Access to information, enabled by the cell phone, also reduces costs. Farmers, for example, don't have to invest as much time as before in discovering price information or weather forecasts. This improves production and productivity.[23]

Enabling entrepreneurship. Another consequence of mobile money and access to information is that cell phones help more people in developing countries start businesses. Entrepreneurship is important for growth because it creates jobs and, often, fosters new innovations. For governments, phone-enabled business creation helps more entrepreneurs enter the ranks of the formal economy. That can boost public revenues. The head of one mobile money company in Ghana says, "Businesses have been established because of mobile money."[24] According to USAID, 41 percent of people use their mobile phones "to increase their income and professional opportunities."[25] The spread of mobile money helps entrepreneurs, too, as it lowers the cost of taking payments—a business doesn't need a credit card machine and its associated fees if a simple transfer between phones can take place.

Closing gender gaps. Globally, extreme poverty appears to be gender-neutral, with a fifty-fifty split between men and women. But in many countries, women face discrimination, fewer opportunities than men, and lack of access to tools they could use to improve their situations. Cell phones are not a silver bullet, but they help close the gap between men and women in many areas. This includes access to financial services and greater consumption of goods and services by women.[26] Mobile money in particular has opened up greater occupational opportunities for women. Financial inclusion and access to financial products allow women to work outside the home and escape the farm.[27] That strengthens their economic role and helps reduce obstacles due to discrimination.

I am an optimist; you've figured that out by now. I'm optimistic about people and about the future of mankind. History bears me out. By virtually every measure, the world is a better place today than it has been in the past. This includes, but is not limited to, the steady downward march of the number of people living in poverty.

Child and maternal mortality have dropped sharply, with the decline especially rapid in emerging markets and developing countries over the last few decades. The share of people suffering from hunger and dying from famine has fallen for years. Meanwhile, literacy and schooling have grown enormously, helping the entire world grow more productive. We create more wealth than ever before even as we work fewer hours and enjoy more leisure time. Humanity has grown healthier, wealthier, and happier than at any time in history.[28]

This enormous progress has constituted humanity's "Great Escape" from hunger, disease, and early death, and it applies, with a few unfortunate exceptions, to nearly every part of the world.[29] How did it happen?

Human ingenuity. Technological innovation. Visions of a better world.

But we are far from finished. There is still too much poverty, too much intolerance, not enough constructive collaboration. We are on the cusp of even more material and social progress. How we learn, our educational system, is being revolutionized. Health care is being transformed. Even the way we communicate—already the subject of major change—is due for a leap in progress.

Yes, I know about the dire predictions of disastrous climate change. Even now, at the time of writing, we are struggling with coronavirus. But I have an abiding belief in humanity and the ability of humans to survive and improve. When we decide to take climate change seriously, scientists, engineers, and even politicians will figure out how society can mitigate climate change or, failing that, accommodate without disastrous consequences.

My optimism about humanity includes a belief that technology will solve the material and environmental problems of society; education will take care of the rest.

There have been few technologies that have promised more and delivered more than the personal, portable, handheld phone. The cell phone has already had a major impact on people everywhere. Yet the cell phone has only been with us for less than forty years—we really are only at the end of the beginning of its true impact.

Based on my optimism and experience, I have a vision of how the world can and *will* be if we allow ourselves to create it. It's also a vision of the world as I want it to be, of an even better world for more people. Neither wishful thinking nor science fiction, my vision is based on science and fact, especially the direction of wireless communications technology and its impact on different areas. Subsequent chapters lay out this vision. I'm sure to be as wrong about some things as I am right about others.

My vision of the technological possibilities of the future is based on observations and insights derived from over six decades of work on wireless communications. These are the "Marty's Maxims" listed earlier, and I've already discussed some of them. Not all readers will agree with those observations nor with the forecasts I make from them—and that's good. These are intended to be thought-provoking and informative. But I firmly believe in the insights on which my vision is based. No matter what aspects of my vision turn out to be wrong, the underlying observations will, I believe, remain true.

Before describing this vision, I need to address government policy. My vision of the technological future and the continued positive impacts of the cell phone depends in large measure on public policy. The government played a large role in shaping the path of portable wireless communications in the second half of the twentieth century—and that role will be similarly large over the next several decades. We won't realize the technological possibilities in front of us if we don't get policy right.

AFFORDABLE AND ACCESSIBLE WIRELESS IS THE PUBLIC INTEREST

I don't fly to Washington, DC, nearly as often as I was required to do in the 1970s and 1980s. But I do still travel there at least four times per year (or, at least, I did before COVID-19). I serve on the FCC's Technological Advisory Council along with fifty or so technologists, mostly representing industry leaders. I also serve as cochair of the working group on antenna systems technology and am a member of the working groups on fifth generation (5G) networks and the Internet of Things (IoT).

Everything has changed about air travel in the last forty years. There are long security lines, planes are more fuel efficient, the food has mostly been reduced to pretzels, there is no smoking allowed onboard, and so on. And, of course, we use our cell phones to make ticket reservations. Yet when I'm in Washington, participating in meetings about spectrum management and telecommunications policy, it often feels like very little has changed. In fact, I find myself addressing some of the same issues as in the early 1970s.

Large telecommunications companies are today trying to persuade the FCC and the public that they need large swaths of spectrum to accommodate the continuing growth of their markets. Once again, claims are made that we have a spectrum availability crunch. And arguments are flying back and forth about what parts of the spectrum should and shouldn't be used for certain uses. Everyone, of course, is seeking to acquire the exclusive right to use more spectrum, and each thinks their needs are the most urgent. Few are willing to share their radio channels—they insist on exclusive use.

This isn't due to stagnation in technological development. The wireless technology we use today is light-years ahead of where it was forty years ago. The reason the policy debate is mired in a time capsule is because we continue to ignore very basic principles of the radio frequency spectrum.

This worries me—and should worry everyone.

If our public policy framework for spectrum management does not catch up to technology and doesn't account for these basic principles, we'll see slower progress in many areas. We won't see the remarkable advances in education, collaborative work, and health care that are at hand and that I describe in the following chapters. We won't realize the full potential of the cell phone's impact on society. The natural resource that is radio frequency spectrum has been called "economic oxygen" for entrepreneurs, large companies, entire industries, and the national economy.[1] Others refer to it as "the shared resource that perhaps most strikingly and most pervasively affects the well-being of society."[2] I see it as the most valuable economic resource of our current century.

Yet the terms "radio frequency spectrum" and "spectrum management policy" don't exactly light up people's eyes. More likely, they make eyes glaze over. The average person doesn't give spectrum, let alone spectrum policy, much thought until her signal is lost or call dropped. That's not a big deal: we don't necessarily need millions of spectrum experts walking around. But I believe everyone can grasp a handful of basic principles that should guide and inform spectrum management over the next several decades. To me, they are fundamental truths that help point us toward a bright future of technological progress.

$$\smile\!\!*\!\!\frown$$

Let's start with the most basic. The radio frequency spectrum—which carries information between cell sites and your phone—is public property. Put another way, we the people own the spectrum and provide bands of it, through licenses, to entities such as cellular operators and television stations. By law, these entities must use their bands of spectrum in ways "consistent with the public interest, convenience, and necessity." That seems pretty straightforward but has led to wide interpretation of just what exactly constitutes the "public interest." Some policy experts call the standard "operationally meaningless."[3] Former FCC chairman Tom Wheeler said he found during his tenure that it was a "pretty malleable concept."[4]

The overriding objective of spectrum management in the twenty-first century should be enabling the provision of ubiquitous and affordable wireless connectivity for everyone. People care less about bandwidth (for example, how long it takes to download a movie) and more about whether they can afford wireless access in the first place. That is the true public interest.

In my decades of interacting with the FCC, I've never known the FCC to satisfy everyone; the notion of "public interest" is continually shifting. The entire monopoly of the Bell System was justified as being in the "public interest." Universal telephone service was achieved. Consumers (the public) had to pay higher prices, but that was a cost everyone had to bear for universal service. Limitations on competition were justified by the "public interest."

Distortions in the name of the public interest have not been confined to telephones. In the 1930s and 1940s, the public interest was invoked to protect local television broadcasters to, first, slow down the development of FM radio and then, later, to move the entire FM band at the behest of television networks.[5] In the 1960s, the FCC helped protect traditional TV networks by denying and revoking licenses for emerging operators in the new business of cable television. The stated "public interest" was in protecting local TV stations and allowing existing broadcasters to keep the public informed. Oh, and cable television would never amount to much of a national market anyway. (Sound familiar?) Cellular telephony was so slow-moving in the 1970s and 1980s in part because of spectrum protection—for the public interest, of course—of existing users.

The FCC has an incredibly tough job, and this is its most challenging task. It is a complex balancing act between users such as emergency responders, entertainment companies, unlicensed spectrum uses (including Wi-Fi), cellular providers, satellite companies, and more. Federal use of the radio spectrum, including military, is managed by the National Telecommunications and Information Administration (NTIA), an agency of the US Department of Commerce. The FCC oversees everything else. The complexity of that task is illustrated in the frequency allocation "wall chart" produced by the NTIA (see next page).[6]

The horizontal bars indicate general activity permitted in a given band: "federal exclusive," "federal/non-federal shared," and "non-federal exclusive." The colored segments denote what radio services are authorized to use the frequency in that band. This includes thirty different services, such as "aeronautical mobile satellite," "broadcasting," and, of course, "mobile." In the early days of radio, the only useful bands were those at the top of the chart: the low frequencies.

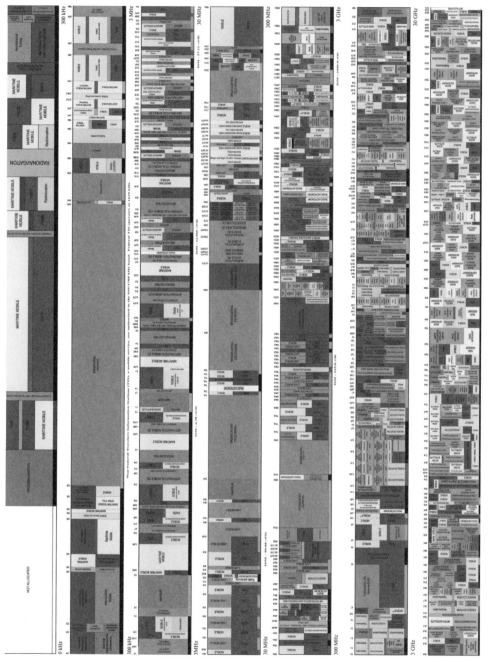

Radio spectrum frequency allocations in the United States.

(Here, the bars on the lower or bottom part of the chart are *higher* frequencies.) As technology continuously advanced, more and more bands at higher frequencies became useful.[7] The initial applications of commercial radio were in shipboard Morse code transmitters. Radio and TV broadcasting occupied most of the usable spectrum for many years, but satellite systems, two-way land mobile radio, cellular, and many other services now compete for access to the spectrum.

The FCC and NTIA do not deliver radio service. Their responsibility is to make sure that the public receives the many benefits made possible through the use of radio spectrum. As merely a quick glance at the NTIA chart makes clear, hundreds of services compete for spectrum. Most of them are considered vital and society cannot function without them. Competition keeps getting fiercer as more and more activities depend on spectrum. Does anyone envy the FCC's task at managing all this?

The agency has come under fire for nearly its entire existence over the public interest and other issues. The many commissioners I've known personally have been dedicated and sincere public servants. Of course, they had, as political appointees, divergent opinions and viewpoints. The decisions with which I am familiar were always made with the intent and belief that they were in the public interest. Our governmental system has maintained a balance at the FCC that has allowed the United States to lead the world in communications policy and technical progress. During the hearings in 1970 and 1973, I especially admired Dick Wiley and Nick Johnson, both in their mid-eighties today and still active in the profession. I was fortunate, in recent times, to befriend Commissioner Robert McDowell, a gentleman and a scholar, who has personalized my understanding of the sacrifices these public servants must make.

When it comes to spectrum management and wireless communications, the FCC should be guided by a notion of the public interest as ubiquitous and affordable wireless connectivity for everyone. Full stop.

In addition to grounding sound policy on the vital public interest of shared spectrum, a good deal of the controversy in managing spectrum use results from a view that spectrum availability is scarce. It's treated like "beachfront property"— when it's occupied it's gone.

The "silent crisis" of spectrum in the 1960s was grounded in this view. Today we are told we are in another crunch of spectrum availability. The proliferation of uses for mobile devices is driving up demand for bandwidth. More and more people are using their phones to watch live sports, television shows, or YouTube videos. Satellite systems are proliferating to bring more affordable broadband access to more people. Additional spectrum is being assigned for 5G networks (and, eventually, a never-ending series of successive "Gs"). New technologies create not just greater demand but arguments over spectrum priority. At the time of writing, a dispute had erupted between the FCC and the Department of Transportation over spectrum use. The FCC proposed repurposing part of the "safety spectrum" for Wi-Fi, while the DOT proposed it be held to permit autonomous cars to connect with each other.

Scarcity has even been legally enshrined by two Supreme Court decisions. In 1943, the Court observed that "the radio spectrum is simply not large enough to accommodate everybody."[8] A quarter-century later, the Court wrote that "broadcast frequencies constituted a scarce resource whose use could be regulated and rationalized by the Government."[9]

The NTIA frequency allocation chart still conveys this scarcity perspective visually. The colored bars of varying size certainly make it look like decades of spectrum fragmentation have left little room for additional capacity.

Yet the "silent crisis" in the 1960s never became an audible one. The NTIA wall chart has grown larger and more colorful as it has accommodated more and more spectrum use. The Supreme Court has been proved wrong on its characterization of spectrum. Today's apparent spectrum shortages will likewise disappear. Why?

There has never, in the history of radio, been a scarcity of radio spectrum, and there never need be a scarcity in the future. Since radio's invention, technologists have created new spectrum faster than uses for that spectrum have been invented. Technological advances in spectrum use and efficiency have continually kept ahead of ever-rising demand. This has been achieved through expansions along both the extensive and intensive margins. We have consistently extended the use of higher and higher frequencies, and we have figured out how to intensify the use of frequency bands.[10]

In doing so, technologists and engineers have continued to prove the enduring truth of the Law of Spectrum Capacity, known to some as Cooper's Law. Technological advances have allowed us to double the capacity of the radio waves every two and a half years since the dawn of radio. Today's spectrum capacity is more than *one trillion* times that of Marconi's day. We already know, at the time of this writing, that we can continue this rate of improvement for another sixty years with *existing* technology.

The limitation on spectrum use is not availability, but our technological capabilities and imaginations in how to efficiently use spectrum. As the President's Council of Advisors on Science and Technology has recognized, spectrum shortage is an "illusion."[11] To others, a "myth."[12] Instead, we are "swimming in underutilized frequency spaces."[13]

There is an abundance, not a shortage, of spectrum.

A mistaken perception about a shortage of radio spectrum has dogged the management of spectrum for its entire history. Public regulators and private companies—and those in many other countries—have created "artificial scarcity."[14] This has created inefficiency, with large swaths of spectrum going underused even as the social and economic potential for use have been clear. And *that* goes against the public interest.

In its approach to regulation and spectrum management, the FCC should trust that technological innovation will continue.

There is still extraordinary inefficiency in the use of spectrum for communications. Transmitters in base stations and cell phones send signal energy in all directions. Only a tiny portion of this energy is useful. The rest is wasted—in fact, it interferes with other communication. With existing smart antenna technology, it's possible to deliver the signal energy directly to a user with little waste, creating, in effect, a personal cell. The result is multiplication of spectrum capacity.

Systems of the future implemented with technology already in development today will have capacity at least millions of times greater than existing cellular systems. These systems will be implemented in networks of nodes, similar to today's cell sites, but controlled by artificial intelligence that continuously optimizes use of the spectrum. The wasteful exclusive allocation of spectrum will no longer exist. Vigorously competitive cellular operators will manage virtual systems that will be optimized for different types of applications focused on making their customers' businesses and lives more efficient and more effective. Embedded artificial intelligence systems will ensure that every device using a node will automatically optimize spectrum use. Time and bandwidth will be minimized, latency will be appropriate to the application, and the optimal frequency band will be used for each transmission. Spectrum sharing will also continue to increase; the FCC is already assigning more spectrum shared between different applications and users. Technology will continue to increase that capacity.

When all of this happens, we will have realized the original promise of the DynaTAC. Dynamic, adaptive, total area coverage.

⌒⋇◯

This scarcity perspective has been reinforced ever since the US Congress first authorized and then encouraged the FCC to auction spectrum licenses. Yet spectrum auctions trade long-term effective use of the radio frequency spectrum for short-term expediency. Auctions of the radio frequency spectrum tend to concentrate owner-ship of spectrum licenses with the wealthiest operators, who usually focus on the most lucrative markets and tend to neglect customers in rural and poor urban areas.

Most spectrum auctions should be abandoned.

For much of the twentieth century, the FCC assigned portions of spectrum through "comparative hearings." Spectrum licensees were selected based upon promises of future performance in delivering services. During the 1980s—prompted in part by the deluge of applications for cellular spectrum licenses—the FCC experimented with lotteries. Then, in the 1990s, auctions became the predominant approach to spectrum assignment.

Auctions were adopted to bring market discipline into spectrum assignment. In contrast with the "beauty contests" that comparative hearings came to be known as, auctions are seen as promoting the highest and best uses of spectrum. A company that sees large economic potential for a certain band will presumably bid high for it and then exploit the band to the fullest to recoup what it paid—hopefully for the benefit of consumers. Auctions also raise billions of dollars in government revenue. The money raised by auctioning spectrum is usually a small fraction of the real value of that spectrum.

Spectrum does not, by itself, have value. It's the information made available by that capacity that has value. This forms part of the rationale for giving private companies an assigned frequency band; they have made a judgment as to its eco-nomic value and are best placed to reap that value by exploiting the capacity.

Spectrum auctions appear to have worked better than what came before, but they have not solved key challenges with spectrum management. For one thing, they almost guarantee that those with the most money will get the most spec-trum. If that were a guarantee of maximum capacity exploitation, it wouldn't necessarily be a problem. But auctions still result in suboptimal use of spectrum because of those persistent notions of exclusive use and artificial scarcity. Our entire regulatory regime has created a situation in which value is seen to be had merely in the possession of spectrum, not full use of its capacity.

What has happened is entirely predictable. Telecommunications companies say they need more and more spectrum—*even as previously allocated spectrum bands are not being fully utilized* to benefit consumers and the public through, for ex-ample, broader coverage and lower costs.

We've told ourselves that spectrum is valuable not because of potential use, but because its capacity is limited. Therefore, we have to assign bands of it in ways that limit interference and respect an auction winner's exclusive use. We assume their bid is a proxy for the minimum economic value they will seek to generate from it.

This approach creates short-term benefits—government revenue, an efficient assignment process, clear rights—but at the expense of long-term benefits. We miss out on the full maximization of spectrum use.

The key to making spectrum management workable and better is competition and after-the-fact oversight. This just doesn't happen today. With auctions, the government finally brought market signals into spectrum management. But improvement is possible and necessary. The government reaps financial benefits through auctions but doesn't ensure that a license holder is using spectrum productively.

If an organization isn't using their spectrum, or isn't using it efficiently, it should be forfeited. This is similar to what is done in the management of mineral resources. I would not, however, resurrect the old "beauty contests" of comparative hearings. There are likely various ways to monitor inefficient use of spectrum. One proposal has been to build on the "finder's preference" program that used to be run by the FCC.[15] Technology can enable us to monitor spectrum utilization, and I have faith we can develop innovative policies to enforce it.

Another interesting idea that has been proposed is that of depreciating licenses. In this scheme, licenses would be perpetual, not in need of periodic renewal. A licensee would pay an annual fee equal to 10 percent of the license value, in effect purchasing that 10 percent from the FCC (the depreciation component). The catch is that the value of the license is self-declared by the licensee—and it serves as a declaration of offer. The license holder must be willing to pay at that value if someone makes an offer.[16]

I don't know if these alternatives would work, but the point is that we need to ensure effective utilization, not hoarding. Auctions were an important step, but we're not quite there yet.

Above all, our public policy framework for spectrum management needs to encourage technology focused on people rather than places and things. This has been a challenge for a century. Today, 5G technology is on everyone's minds and mouths. It figures centrally in global geopolitical affairs and the strategic calculations of large telecommunications companies. It presents a number of economic and social possibilities.

Those include: Better management of utilities and resources. Increased public safety through faster data transmission from surveillance devices. Ambulance drones, robot surgeons, unprecedented integration of medical data sets. Predictive maintenance of farm equipment. Shopping with augmented and virtual reality, facilitated by automated warehouses. Downloaded movies to your device in seconds. A connected cloud of machines everywhere.

This is the enticing technological utopia, the Internet of Things, promised by 5G and touted by the telecommunications companies who want to bring it to you.[17] To make this utopia a reality, however, those companies have taken the position that they need additional, exclusive access to more spectrum.

But most of the technological opportunities and underlying technical needs can be served by existing spectrum. That's because of continuing inefficiencies in how existing spectrum capacity is used. The Internet of Things makes for great marketing material, but the companies pushing it have neglected to create an Internet of People.

The best way to do that is to promote competition in wireless communications—competition that encourages companies to prioritize the needs and interests of people. Today we find that history is repeating itself. Dominant near monopolies in telecommunications are saying they must have additional large chunks of the spectrum, a publicly owned resource. They are not yet, however, using available technology to expand the capacity of their existing spectrum and increase wireless coverage and reduce prices so that more people can enjoy the benefits of connectivity. Cost and coverage are the reasons that millions of Americans still have inadequate access to necessary wireless connectivity.

Tension between protecting monopoly and promoting competition has marked telecommunications policy from the very beginning. AT&T's slogan during its twentieth-century heyday was "The System is the Solution."[18] When television first began to develop in the 1920s, it was a threat to the large established radio companies (such as RCA) and their networks (RCA owned NBC). The head of RCA, David Sarnoff, "repeatedly lobbied the FCC to adopt their view that television was simply an outgrowth of radio, and one that only the established radio industry could be entrusted to bring to proper fruition."[19] An RCA submission to the FCC stated, "Only an experienced and responsible organization such as the Radio Corporation of America should be granted licenses to broadcast material, for only such organizations can be depended upon to uphold high ideals of service."[20] That should sound familiar.

Large companies can compete ruthlessly with one another and with younger and smaller companies. In the 1950s and 1960s, Motorola effectively operated as a monopoly in two-way radio communications, by virtue of its superior technology and marketing. We were not, however, a government-sanctioned and protected monopoly. Nor were we without competitors—we lost contracts to American and Japanese companies.

Competition, not consolidation, is what produces better outcomes for everyone. International comparisons find that countries with more competition in telecommunications have lower prices for consumers—without any loss of service quality.[21] We cannot let monopolies, or near monopolies, define our technological future for us.

I know for certain that we have not reached the end in terms of spectrum management. Room for improvement, like spectrum capacity, is plentiful. It requires policy makers to trust in technological innovation, promote competition, and establish people's access to affordable and ubiquitous wireless connectivity as the ultimate purpose of spectrum policy.

TRANSFORMING
HOW WE LEARN

At my elementary school in the 1930s, a consistent feature of every classroom was a leather strap that hung threateningly from the teacher's desk. Students who got into fights or sassed the teacher painfully discovered its purpose.

Mercifully, leather straps have disappeared. But there is a new consistent feature of classrooms, one that is the scourge of teachers. That's the cell phone. In the United States, most high schoolers have one, over half of middle schoolers do, and even about 40 percent of elementary school students have a cell phone.[1] They're not being used exclusively to text during class. Students use their cell phones to collaborate on group projects, see new assignments posted by the teacher, check their grades, and more.

This is just one example of how much wireless technology is influencing K-12 education in this country and elsewhere—and how far we still must go. Implementation of more technology in classrooms has long been a stated goal of educators and policy makers. Successive presidential administrations have promoted broadband deployment in schools, large technology companies have made software and hardware available freely or cheaply to schools, and numerous philanthropic initiatives have promoted education technology. Many schools are using technology to "flip" the classroom, with content delivery done outside of school. Traditional homework is done in the classroom, with teachers acting as advisors catering to the specific needs of individual students.

Yet the promise of technology's impact in the classrooms has often been more hype than reality. One of the biggest challenges is access to the necessary technology. Millions of American students have limited access to the internet or any form of wireless connectivity. Comparable numbers of students cannot afford access even when it's available. Only 38 percent of US school districts meet the FCC's declared internet speed for digital learning of one Mbps, or megabits per second.[2] The median bandwidth speed among schools is 676 Kbps, just over half the FCC standard.[3]

This is part of what's called the "digital divide" in the United States, with unequal access to wireless connectivity determined by race, income, and geography.

- Black and Latino households have lower access to home broadband than whites. For Latinos, 61 percent have home broadband; for Blacks, 66 percent have home broadband. These compare to 79 percent for white households.[4]
- In rural areas, 63 percent of Americans have home broadband, compared to 75 percent in urban areas and 79 percent in suburban areas.[5]
- Nearly one in five Black and Latino students (19 and 17 percent respectively), have *no* internet access (or dial-up only) at home.
- One-quarter of students living in households below the poverty line have no internet access (or dial-up only) at home.[6]

But they do all have cell phones. Cell phone ownership is the same among Black, Latino, and white populations at over 96 percent. It's above 92 percent for all levels of education and above 95 percent for all income levels. And cell phone ownership is over 90 percent for all age groups and all geographic areas.[7]

Those cell phones can and should be the basis for an entirely new infrastructure to close the digital divide. Our education system is important enough, and broken enough, to justify creation of a dedicated and independent wireless broadband system that can bring affordable connectivity to every K-12 student. The freedom of affordable wireless communication should be made available to every student. And students in high schools and below should be restricted to a search engine that is designed specifically for education. I call this vision "Education 2.0."[8]

Halfway measures will simply not work. We are leaving students who can't afford wireless access behind. We are leaving behind students in rural areas and in low-income urban areas that have no coverage. We need to not only close the digital divide but also ensure that learning can and does happen *anywhere* for *everyone*.

Technology, however, is just the beginning.

⟡

When I first started college at the Illinois Institute of Technology, I began studying physics. One of the requirements for a physics degree was to take chemistry, which included a chemistry lab. As some readers know, passing a chem lab requires discipline and a grasp of theory.

I never quite mastered it. In any lab experiment, I would inadvertently discard the wrong elements or compounds and get the wrong result. The second year of physics required even more chemistry—there was no way I was going to survive that, so I switched to electrical engineering.

Engineering thrilled me because of its orientation toward tackling real-world problems. You still need some understanding of how and why certain things work or don't—how a transistor functions, for example. But tinkering with things comes first. The world needs theory, of course; the entire notion of cellular service was dreamed up as something theoretical. But I've always been drawn to the doing, to just figuring out how to make something happen without first stopping to understand the why.

Part of this was my own education at a technical high school, a technical university, and then the navy. It was all about solving real-world problems—*doing*, in contrast with passively sitting in a classroom listening to a lecture. It was all about bringing multiple disciplines to bear on those problems, in contrast with focusing on silos of learning like math, language, and geography. Students motivated to solve problems will learn faster and better than those who are simply told or taught to memorize something.

This approach, as far as I can tell, has been catching on in the education system. Many schools at every level have recognized that the traditional lecture is suboptimal. They are confronting their students with real-world problems, asking them to apply experience and knowledge, thereby continually expanding that experience and knowledge. Whether it's Canyon Crest Academy, a high school near my city of Del Mar, or Illinois Tech, more schools see this as the basis for more effective learning.

I offer two examples of teaching methods that are replacing the lecture in progressive schools—electronic gaming and project-based learning (PBL).

Consider the nature of successful computer games. They are engaging, interesting, entertaining, and addictive. They are also adaptive; a player who lags gets to repeat or take an easier route. And when a person finishes a game, she has already been tested to the limits of her ability; no need for a grade. Doesn't this sound like an ideal educational platform?

An encouraging trend in educational technology is development of games that teach people, challenge them, measure their progress, and simultaneously entertain them. The curriculum then becomes a diversity of appropriate subjects. Tests that measure progress are embedded in the games. It's almost impossible to cheat because the game catches it. In this model, education happens everywhere and all the time, not just in school. Much progress remains, though, in making these games true combinations of both education and entertainment.

A second example is PBL. Project-based learning has gained more support and adoption in recent years. The idea is implied by the name. Students work together or alone on projects (sometimes called challenges) that require them to bring knowledge from many subject areas to bear.

Let's say the challenge is to improve the performance of a baseball bat hitting a ball. First, students will play around with the bat and ball to get a feel for how they work at different angles and speed. They'll then need to understand the physics of bat on ball. Digging into the engineering of the bat and ball is also required. The next part of the assignment requires them to present their solution as if they were selling it to a potential customer. They need to bring in some economics and business to get pricing right. The students must develop a presentation that pushes them to hone their verbal communications skills while also understanding the most effective visual graphics that will enhance their sales pitch. All the while, they're using various computer programs to assist them, gaining experience with those tools.

And, of course, the students should have enjoyed consistent wireless connectivity, at school and home, throughout the duration of the project.

This is not just about science and engineering. A project-based learning approach applies to the arts as well. Consider music education. A grasp of musical theory and composition structure is helpful, but what's the first thing done in elementary school music classes? Fiddling with the instrument. Tinkering goes along with the teaching of theory; it's not subordinated to it.

⌒✳⌒

I am by no means an expert on education. But I've spent the last seven decades learning in many different contexts. My formal education—at Crane, IIT, and in the navy—was a critical foundation for that learning. Through my years at Motorola and across the various companies I've helped start and grow, that learning never stopped. Based on these experiences, I know for certain that two things will transform our education system for the better: Affordable wireless connectivity for everyone, everywhere. Real-world problem solving.

If we close gaps in access to wireless connectivity, we expand the scope of who is participating and learning. If we challenge them to solve problems, we expand the possibilities of what technology enables us to do.

This approach also requires improvements in how people learn to work together. That is the subject of the next chapter.

ENHANCING HOW WE WORK TOGETHER

S omeday, not long in the future, you will be in the shower. A novel idea will strike you—as they so often seem to do in the shower. You'll start talking out loud, seemingly to yourself, about the idea. But your audience is your cell phone.

The mobile device, upon hearing your voice, will store the idea in a searchable format. It will analyze the content to extract key issues and tags. The phone will identify which members of your team the content is relevant for and distribute your idea to them with the appropriate level of urgency. Your phone will also respond to you.

"I've stored your comment. It looks like you are referring to the Omega project. Shall I post your idea to the group's bulletin board?" The phone will continue, "I have searched the concept you just described and find reference to it in several company sites, all of which are now in your Omega folder. I also noted that some original work in this area was done by an engineer from our company in Germany. You may wish to invite him into the Omega group."

By the time your shower is over, other members of the Omega group have weighed in on the idea and the research has been distributed to all those user devices in the collaboration group that may find your thought useful.

Welcome to the future of collaboration, in which your cell phone becomes a true platform enabling you to work together seamlessly with others. Technology today already provides various ways to do this. Google Docs, WhatsApp, Twitter, Slack, Asana—the list of collaborative tools keeps growing. The COVID-19 pandemic has taught us the necessity of still fairly new video platforms such as Zoom, Teams, Google Hangouts and Meet, BlueJeans, and more.

The heavy use of these video platforms during the pandemic has also reinforced the importance of face-to-face interaction for making collaborative work possible and effective. Email and messaging apps are must-haves but are no substitute for talking directly to someone. For most of my career, the need for face-to-face interaction led to inefficiencies. It meant a lot of time spent on airplanes. That has continued to be the case for many people today. COVID-19 and the improvements in video platforms may mark a new era in remote collaboration.

Yet as my crude shower example demonstrates, we are still only at the earliest stages of collaborative technology enabled by wireless communications. (And there are limits, of course—few people will opt for video meetings while they shower.) While technology continues to give us new ways of interacting with each other, we also need to address other challenges. We need to figure out how to get out of the boxes we keep creating for ourselves and how to get along better in teams.

Think outside the box. This is by now a well-worn, clichéd phrase found more often on semi-inspirational corporate posters than as serious guidance for collaborative work.

But we still haven't figured out how to actually get outside the box. There is a constant tendency—in companies, universities, nonprofits, governments—to draw lines and boxes around ourselves. We insist on categorizing our problem-solving processes, knowledge searches, and teaching methods into narrow, self-contained silos. We keep inventing new words to try to get ourselves out of these confinements: multidisciplinary, interdisciplinary, cross-functional, cross-pollinate, and so on.

Creating lines and boxes is a natural impulse of all organizations; it's why they exist in the first place. But new creations and solutions to problems tend to draw on many different specialties and mental models from multiple disciplines. The best way to have people think outside the box is to not create the box in the first place.[1]

During the process of creating the DynaTAC in 1972 and 1973, I was able to bring together people from multiple areas by either ignoring organizational boxes or simply barging through them. I was good at conveying my enthusiasm for new ideas to others, which helped overcome barriers they may have faced to working with me. I encouraged the team to visit with customers, learn about their experiences and complaints, and use this feedback to refine product designs.

The very idea for the handheld, portable phone was devised beyond the box. I constantly wandered around the Motorola labs and groups, learning from everyone with whom I interacted, introducing them to others who had complementary skills and ideas. This familiarized me with where various pieces of technology were in the developmental process and what might be possible to bring together in a new device. Envisioning the cell phone wasn't a matter of sitting down and theorizing about new means of communicating. It was the result of a logical process of knowing where technology was headed and orienting that direction toward the notions of portability and meeting the needs of mobile people.

The challenge for any organization is in making this type of process, which can seem ad hoc, something more formal and intentional—without creating new boxes. The heart of that challenge is forming productive teams. Groups of individuals—i.e., teams—have always been important to innovation. In any field, what business guru Warren Bennis called "Great Groups" have been on the front lines of collaborative and creative problem-solving.[2] Our DynaTAC team was a Great Group. The fact that the group had a common, well-understood goal made each member greater.

How can collaborative groups be assembled based on the ability of participants to contribute rather than their position on an organization chart? The greatest strength of a successful team is diversity—diversity in skills, experience, gender, ethnicity, age. Yet assembling diverse teams continues to be a challenge for every type of organization. We are inefficient at communicating within and between organizations and at removing the boundaries of time, location, and calendar.

Collaboration traditionally revolves mostly around meetings, both physical and virtual. Let's consider the example of an executive tasked with solving a major problem or making a critical decision. Her first step is selecting invitees. The traditional source is the organization chart, which constrains us through the intimidating tradition of not inviting people too many levels separated. A meeting location needs to be selected, and then a time when every participant can be available. When the group is finally assembled, the leader announces, "That's it, team, we've got an hour to solve our problem." The team members then start diverging the moment they leave the meeting.

Every team needs to continually refresh collaboration. The best way to do that continues to be more face-to-face meetings. Which is inefficient and time-consuming, especially when teams are widely distributed, even within the same building.

We can do better. Although many of us have become skilled at using social media for fun and specific tasks, most collaboration today is suboptimal. The right people are not exchanging ideas often enough about problems and decisions that require action—just at the time that problems and decisions require better collaboration! We have the tools. We already talk, text, tweet, post, email, and videoconference on our phones—mostly inexpensively, and mostly conveniently—but we haven't yet developed the new habits and processes to use them more collaboratively. That's about to change. Collaborative tools will enhance the quality and success of teamwork by improving the quality of our people networks and the quality of communications within those networks.

～#～

Dr. Harinath Garudadri, a research scientist at UCSD, is pushing the boundaries of current collaboration tools in pursuit of a solution for hearing loss.

"People who hear well are happier, experience more of life, and live longer," Hari says. "The opposite is true for people with hearing loss. We have the science of hearing that can help those with poor hearing live normal lives."

I was introduced to Hari, a genius whose passion is bringing that science to practical reality, by his department head, Professor Rajesh Gupta. Rajesh and I have firsthand experience with hearing deficiency.

Of all the human senses, hearing might be the least appreciated. That's because the hearing loss that *everybody* experiences as they age happens gradually. It's so gradual that people are unaware of what they're missing until they become isolated from their colleagues and loved ones. The isolation typically starts with one of the following: "What?" "I'm sorry, could you say that again?" "Did you say something?" Eventually, one reaches the point of embarrassment at having to ask and can no longer converse comfortably.

Most people have adequate hearing until they get older and then, too readily, accept hearing loss as an inevitable result of aging. For some, the loss of hearing can be devastating. As their hearing deteriorates, they find it harder to engage socially with others, especially in noisy restaurants and in lecture halls, classrooms and theaters, where reverberations make it impossible to understand voices. Hearing aids can be helpful, but, aside from the ridiculously high cost, there is a social stigma associated with wearing them.

Most of all, hearing aids don't work very well. As a result, the hard of hearing, which includes almost everybody at some time in their lives, gradually opt out of society and often become unproductive, frustrated, and unhappy. Many suffer from extreme loneliness—even though they still have wisdom to share with the world.

Hari, Rajesh, and I categorically reject the idea that the poor hearing of a significant part of our population should result in their being tossed into the trash pile even though they have a correctable problem. We are convinced that technology can restore hearing to youthful levels, and that this can be done at far lower cost than today's unsatisfactory solutions.

To create a solution, Hari and Rajesh have gathered experts in a half dozen disparate fields. One of his most difficult tasks is to get them all to understand and support the breadth of his project and, because funding is so scarce, do it all with extraordinary efficiency.

Aside from this core team of students and researchers, they need expertise and research in psychology, design of integrated circuits, subminiature speakers and microphones, brainwave sensors, and manufacturing of devices as intricate as mechanical watches. They need audiologists to work with ethnographers and both to work in turn with technologists. Complicating matters is the fact that each of these silos of expertise has its own linguistic argot and its own idioms. There is the potential for an unproductive Tower of Babel.

The scope and complexity of collaboration needed to make the hearing project successful would have been impossible to achieve before the cell phone and the internet—together—existed. According to Rajesh, mobile technology permits and enhances collaboration not just based on specialized knowledge or skills—which has existed for thousands of years—but across lines of knowledge and expertise.

The cell phone, he says, "makes boundaries permeable" across disciplines.

This diagram is Hari's way of describing the collaborative team. Each of the letters *A* through *H* describes an expertise or group of experts, as defined in the lowest two layers, who are essential to his project. Each of the boxes is a distinct technological discipline. Few of the people represented by these boxes work for Hari. He needs to share his passion and inspire the various groups to participate in his mission. And he must get them to talk to each other.

Using what they call an "ecological momentary assessment," or EMA, based entirely on mobile devices, Hari and Rajesh are able to overcome the boundaries created by language and knowledge specialization. With the EMA running on everyone's phones—and providing hearing aid wearers with constant feedback—the scientists can see information coming in, communicate together, and collaborate on next steps.

A new kind of collaboration is happening, building on our society's increasing comfort with communicating and sharing via social media. In addition, the internet hosts hundreds of forums, blogs, podcasts, and other ways of people exchanging ideas and wisdom in packets of a few words or lengthy publications or video tutorials.

That is the essence of why Hari and Rajesh's project works. It is typical of the many complex technologies that are being created in laboratories and workshops

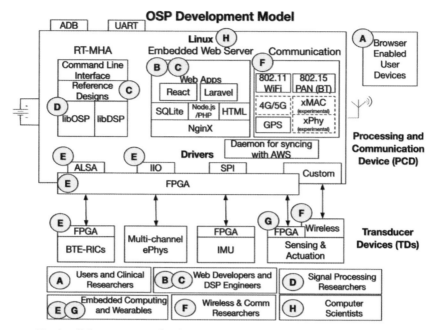

Hari's collaborative team. If you're not a techie, ignore the detail inside the boxes.
They are merely different types of people and technologies.

worldwide that are going to eliminate not only hearing loss but also poverty, illiteracy, and disease.

To collaborate effectively, people need to communicate effortlessly, efficiently, frequently, and affordably. A tall order. Systems are evolving that combine the attributes of social media and other collaborative applications and that optimize communications for specific objectives. But all collaborative systems rely on means by which people connect—wireless communication is the best way to connect people.

The hearing project is still in its early stages. They are shaping the collaboration tools as they work. I'm confident they are headed in the right direction.

Many years from now, we will look back on the early twenty-first century as a primitive time in terms of collaborative tools. We're certainly ahead of where we were when the first cell phones were brought to market. But we will soon find that our ability to get outside of our boxes and work innovatively in teams will be greatly enhanced by wireless technology.

FROM SICK CARE TO HEALTH CARE TO HUMAN 2.0

D id you know that every disease is actionably preventable?"
    This question pulled my wandering attention from the glorious sunset over the Pacific Ocean and brought it back into the dining room of the Lodge at Torrey Pines. Susan Topol, a physician, was speaking.

The human body is a biological marvel. It's an amazingly intricate and balanced set of systems. It is also, biologically speaking, a mess. We're loaded with viruses, bad bacteria, toxins, and mutated or otherwise corrupted cells. When our immune system and other repair-or-replace mechanisms are working, they keep these bad elements in check. We are thus defined as "healthy." When we exhibit any of myriad possible symptoms, we describe our bodies as "diseased." For those moments, we have medical providers, hospitals, and pharmaceuticals. That's the health care system.

"We don't have a health care system; we have a sick care system."

This second striking statement came from Susan's husband, Eric, a cardiologist and professor who founded and directs the Scripps Research Translational Institute.[1]

Eric continued, "We wait for people to get sick and then we treat them." Modern medicine, for the most part, treats these symptoms rather than their sources.

But if there was a way to anticipate the failure of our immune system, to sense when a virus or a cancer cell gains control (or gets out of control), we could zap the virus or cancer cell and stop the disease from happening. Sooner or later, according to Eric and Susan, this is going to happen.

They see a very different and brighter medical future than what exists today. They have taught me about where technology is taking us in how we care for ourselves and how we prevent and treat ailments. My conversations with them have enabled me to see wireless technology, with the cell phone at the core, as the crucial weapon in humanity's battle with disease.[2]

The continued evolution of the cell phone will help us conquer disease—and take humans even farther down the evolutionary road.

In the United States, we sorely need a revolution. For three decades we have been debating how to cover more people with health insurance to expand access. That debate will surely continue. But expanded insurance coverage will mean little if all we're doing is increasing access to a sick care system. Mobile health—the application and use of cell phones in health care and medicine—is improving diagnosis, facilitating better treatment, and widening the scope of what care can be provided.

Mobile technology allows us to apply to health care another of the principles I have distilled from my decades of work: customization is the inexorable direction of products and services. This has two implications. First, a product or service optimized for one individual will be suboptimal for every other individual. Second, a product or service that purports to do all things for all people will, most frequently, do none of them optimally.

It doesn't take much reflection to see that our current "sick care" system operates in the opposite manner. The general approach is based on a notion that there is such a thing as a "normal" patient with "typical" symptoms and that their illnesses can be dealt with in the "usual" ways. Recently, there's been growing recognition of the huge variety among humans in how our bodies respond to different treatments. This has given rise to the movement for personalized medicine, which has received more attention and resources. The United States, for example, announced the Precision Medicine Initiative in 2015; the All of Us Research Program at the National Institutes of Health seeks to sequence millions of genomes. Research increasingly highlights the benefits of personalized medicine across a range of ailments.[3]

The cell phone supercharges the move toward personalized medicine: more data collection through cell phones means more personalization and more precision. And that means better treatment and better patient outcomes. Radio waves will connect the phone to sensors on a person's body and other devices, which will be selected and designed on the basis of the individual's genome and medical history. The cell phone will cease to be a universal device that tries to do all things for all people—it will evolve, through data and feedback, into a specialized device unique to its owner.

⌒⋇⌒

Neither characterization, "healthy" nor "diseased," is absolute. A person may be infected with coronavirus, for example, for days before she feels a sore throat and starts coughing—typical symptoms of the onset of a common cold. Cancer cells can multiply in an otherwise healthy body for years before showing any overt evidence of their existence. The threshold between healthy and diseased doesn't really exist; it's a continuum.

To further complicate things, the effort of the repair systems in the body to work on a specific escalating issue, like a virus that is inhaled, dilutes the ability of these mechanisms to manage other health issues that are otherwise benign. The immune system becomes less effective at keeping these other issues in check—it gets "overloaded."

The best that modern medicine can do in detecting these nascent diseases is the annual physical examination. Insofar as these "physicals" discipline people to see a provider who can use the exams to establish a baseline picture of their health, they can be useful. For the purpose of anticipating diseases early enough to stop them, they are almost worthless.

We humans come in all shapes, sizes, and forms: "we're each a one-of-a-kind intricacy."[4] Our vital signs reflect this variety. A low pulse rate can be a sign of excellent health for one person and a danger signal for another. A particular blood pressure reading can be perfectly normal for one individual and life threatening for another. Our DNA programs our susceptibility to undesirable elements in our bodies and environment differently. When a doctor measures your vital signs, your weight, and myriad other diagnostic metrics, she compares those metrics with a range of what we call "normal." Yet these normal ranges are simply too wide to be useful in describing the early onset of disease in a single individual. To do this properly, we need a comprehensive picture of what is "normal" for that specific person, not for some supposedly "average" individual.

That's where the cell phone enters. If you use a modern smartphone and accompanying wristwatch, they know your vital signs, how much you exercise (or don't), and how much you sleep. The latest iPhone notifies you of irregular heart rhythms and heart rate variability. It knows your body mass index because your scale keeps it informed. We're still in early days, but it's clear that your cell phone is establishing a record of what is normal for you, not merely an average avatar. When this information is added to your medical record (an entirely separate challenge too complex to cover here but that will be solved in the coming years), there is an opportunity to turn the physical examination into something entirely different.

If annual physical examinations are ineffectual, what can we do to anticipate the onset of disease and stop it? If some diseases start in a matter of minutes or hours, of what use is an annual visit to a doctor? Well, suppose that we can attach one or more sensors to your body that give you a diagnosis not once a year, but as often as once a minute. And suppose we can deliver that information to a computer that continuously analyzes your results and compares them with what is normal for *you*. Further, suppose the computer can alert a provider, or you, when your measurements move into your danger zone.

Suppose these sensors that measure your vital signs are sensitive enough to detect the onset of diabetes, or a few cancer cells, or the beginnings of inflammation, before any part of your body was meaningfully affected. What if the data from that exam could be correlated with your history, your genome, and a database comprising the histories of hundreds of millions of other people? If we could do these things, treatment of diseases could be targeted and precise and far less invasive than if the disease progressed to the point that the person was finally aware of it.

These supposed situations are already developing.

Cell phone cameras are helping check, for example, someone's throat. Apps are using phone extensions and plug-in devices to take direct biological measurements. This is precisely what the cell phone enables. It brings a new frequency of micromeasurement to health care. Medical apps are using sensors on the body, some seeking out a specific disease. Others, like the Apple and Samsung watches, deliver more general clues to health issues. Already, thousands of people have radio-frequency identification chips embedded in their bodies.[5] That may sound extreme to many people, but consider how short the jump really is between today's activity trackers and implanted devices.

In 2018, there were fifty-two million users of wearable devices in the United States, capturing and tracking information on a variety of things. That was more than double the number of wearable users in 2014.[6] More and more US consumers are using these wearables to track their health and enhance their fitness. The cell phone is becoming the equivalent of a medical computer server. It won't be long before your genetic information is encoded in your phone and linked to the constant monitoring and measuring. Remember the use of cell phones to establish personal identity in Africa and among refugees?

When necessary, your phone will send you reminders and warnings. Eventually, wearable devices—on your body and linked with your phone—will be sensitive enough to detect and anticipate specific diseases. You may not even need something physical on your body: wireless devices could track your vital signs as long as you're in the same room.

Data collection through cell phones and connected sensors will allow the enhancement of provider-patient interaction. A health-care provider—physician, nurse practitioner, physician assistant, nurse—will not need to spend so much time on basic tasks such as checking vitals, data entry, and more. Why will we need most office visits anymore? Some illnesses and procedures will still require an office or hospital visit. But for many, mobile health dramatically lowers barriers to access by saving a visit to a provider.

What about those who do not have access to such advanced technology? The cell phone is already making important contributions in remote diagnostics. Here are some examples:

- Eye doctors in Mexico City are diagnosing people in remote and poor villages using a simple cell phone attachment that costs a dollar or two.
- Pregnant women in Tijuana are undergoing ultrasound examinations while at home being conducted by remote doctors using a simple cell phone attachment.
- A device the size of a stick-on bandage worn by a person who is subject to congestive heart failure detects when a heart attack is imminent, providing time, before the effects of the attack are evident, during which a simple procedure entirely avoids the attack.
- An app uses the accelerometer in a cell phone to detect the unique hand movements that presage the onset of Parkinson's.[7]
- In thirty seconds, an Apple Watch can take an electrocardiogram that detects your sinus rhythm, electrical impulses that cause your heart to contract and pump blood out to your lungs and body.[8]

Money is pouring into a burgeoning digital health industry. As of September 2019, there were thirty-seven digital health "unicorns"—private companies worth at least $1 billion—valued by their investors at a combined $92 billion.[9]

There are low-cost add-on appliances for phones that take sonograms, measure glucose, and look into eyes and ears and transmit the results to remote health care providers. Mobile health will bring "concierge" medicine, where a dedicated health care provider is more or less "on call," to more people.[10] There are many digital health startups today working to establish virtual care clinics to better connect providers and patients. Eighty percent of US hospitals say they have adopted some form of telehealth system.[11] Teladoc, offering telemedicine and virtual visits, had nearly three million telehealth visits during the first three quarters of 2019, and now counts thirty-five million members globally.[12] And that was before COVID-19 sent telemedicine use skyrocketing even further.

Reducing the need for physical visits means migration from a health care system based on connected places to one based on connected people. This will help the United States (and other countries) manage one of the largest looming health challenges: an aging population. Today, spending on medicine and health care doubles between ages seventy and ninety—spending in the last year of life alone accounts for nearly 7 percent of all US health care spending. Expenditures on medicine and health for the elderly are also highly skewed, with the top 5 percent of health care expenditures for the elderly comprising over one-third of total health care spending in the United States. As the elderly population grows, these numbers will only rise.[13]

Mobile health offers a way to ease these costs and improve care quality. Data collected by cell phones and sensors—and shared with a provider—should enable the elderly to stay in their homes longer. This will reduce costs and enhance well-being. Chronic disease management, basic screening, and medication reminders are also facilitated by mobile health.[14] These and other uses are not just nice to have. They're absolutely necessary if we're going to manage an older population with larger health needs.

The impact of mobile health might be even greater in developing countries. There, digital health identification through cell phones has already enabled better and more consistent care as well as improved data management. Through phone biometrics, companies are creating access to care that didn't previously exist and connecting different digital health platforms that have not been interoperable.[15] A review of mobile health practices by frontline health workers in developing countries found that cell phones enhanced data collection and tracking, communication between providers, and decision-making about what actions to take.[16] The application of cell phones in health care doesn't just improve the patient experience—it also empowers providers to do their jobs more effectively.

Technology alone will not conquer disease—that will also require a completely new approach to how health care institutions are managed, a new relationship between patients and doctors, and fundamental changes in the industry and how it is regulated. But it's clear that wireless technology will have a very important role. I have great confidence that the promise of the end of disease will be fulfilled.

Think about the onset of a cancer. Detection is the hard part. Once we know where the few cancer cells are, we'll develop the ability to zap them with a laser beam without affecting surrounding tissue. We pretty much know how to do the latter. But modern medical devices can't detect a few cancer cells in a human body—not yet. The sensitivity of these devices is rapidly improving. Within a generation or two, wearable devices will be available that will be sensitive enough to anticipate most diseases. The data accumulated by these devices will be transmitted by the descendants of the personal mobile phone to appropriate computers and doctors, who will, in turn, help stop these diseases from happening.

Additional steps in mobile health are already emerging. Cell phones are helping drive participation in clinical trials. Apple, for example, has rolled out ResearchKit and CareKit, which connect consumers to studies and trials while also providing them with benefits.

The necessary key technologies for sensing disease and sending their descriptive data to remote computers are present or evolving. Powerful computers, like IBM's Watson, can already keep track of medical advances far better than any human. While today's science is about six orders of magnitude away from the ability to detect a few cancer cells, scientists are zeroing in on diseases like congestive heart failure, obesity, and diabetes.[17]

The wireless connections needed to accomplish all of the above and more are already here, and technology will drive the cost of these connections down to levels that will make anticipative health care available to all. But the evolution of wireless technology and cell phones won't stop there. Advances in mobile health will help create something that much more closely resembles science fiction: what I call "Human 2.0."

"Soon, we will have a hearing aid that 'sees' more than it hears." That, according to Rajesh, is where the hearing aid he and Hari and their team are developing is headed. It will become a "cognitive assistant," reading facial expressions and nonverbal signals from others, communicating those to the wearer. All the while, the user can adjust hearing aid parameters with their phone—sending data to researchers and clinicians who use it to improve performance.

As your cell phone collects data and learns about each individual user, it will become much more than a phone, more than a smartphone, more than the super-computer in your pocket. It will become that cognitive assistant—an augment.

Ray Kurzweil famously predicted that, by 2045, human and artificial intelli-gence (AI) will merge in the "Singularity." I admire Ray's predictive skills, but I'm skeptical about that prediction. As I said to him in a recent conversation, "A major challenge in the process of amalgamation is the bandwidth of communication be-tween human and AI. Both the human nervous system and AI are processing at enormous speeds." Achieving the level of cooperation that he predicts will require bandwidths way beyond what can be foreseen even twenty-five years from now.

Ray's response: "There's a fellow named Cooper who has a law that takes care of that." Touché! I fear that neither of us will be present in 2045 (although we are both trying hard to do that) to resolve our disagreement.

Generations will pass before an artificial intelligence will be created that is in any way comparable to the human mind. It is unlikely that human beings, alone, can ever create such an intelligence. We humans are still groping around the out-skirts of understanding how the human nervous system—especially the brain—works. So, when an intelligence superior to the human mind is created, it will most likely be created by amalgams of human and artificial intelligence.[18] The smartphone is a primitive beginning to the process of creating such an intelligent amalgam.

Many generations from now, an amalgam comprising a human and an AI aug-ment will achieve greater processing power, higher ability to abstract, and cre-ative abilities far beyond the ability of the human mind to even imagine, never mind match.

Stephen Hawking, Bill Gates, and Elon Musk have chimed in with warnings about the potential danger to society of robotic artificial intelligences that are su-perior to humans and that have no obligation to tolerate, let alone serve, humanity. It is presumptuous of me to take on these illustrious figures, but I don't agree with them. I believe the robots are the next evolution of mankind. The robots are us!

Here's how I expect the evolution to occur.

Today, mobile phone manufacturers and aftermarket developers are trying to get the mobile phone to anticipate the needs of its owner, to act as an assistant. Samsung calls that hypothetical assistant Bixby, a name inferential to a British butler. But there is nothing very intelligent about Bixby, or Siri, or Alexa. To be truly called intelligent, the assistant needs to have the ability to learn, to abstract, and most importantly, to evolve. In health care, this will happen as these non-intelligent assistants gradually become "virtual medical coaches" that provide not only knowledge but also feedback and advice.[19] Hari and Rajesh's hearing aid is one step in this evolution.

The modern smartphone contains, and has access to, a huge amount of information about you, even beyond your contacts, communications, calendar, location, and movements. It can pretty much tell whether you're sleeping or awake, whether you're calm or under stress, and even whether you're alone or engaged with other people. The phone can even figure out, if it is appropriately programmed, whether the people with whom you are engaged are important, as in a crucial business meeting, or casual, as at a family dinner. It would then have a rationale about whether to interrupt or take a message. These examples are only preprogrammed simulations of intelligent behavior.

Over time, the cell phone will transform into an alter ego, an intelligent augment that, acting alone, can make decisions about whether to interrupt you with a phone call or to let you know that an important message has arrived. It will remember for you and make your memories far more accessible than by your fallible wet brain. When your phone learns to figure out what is important to you without being preprogrammed, it truly becomes an extension of you. The combination of you and the phone (if we still call it that) is Human 2.0.

Human 2.0 is more efficient, more effective, more productive, more confident than any existing human, and hopefully, it (or she or he or they) is more relaxed and less stressed. Human 2.0 doesn't struggle to remember trivia; it remembers everything, indefinitely, and provides its human partner with access to all that information. The human partner may now focus on judgment, abstraction, creation—the attributes that make us all human.

You and your smartphone are a primitive version of Human 2.0, but your partnership is evolving. Add smart watches and your augment now knows your pulse rate and skin temperature and has continuous knowledge of your electrocardiogram, your ECG. This knowledge has implications far beyond the obvious ones of evaluating your health on a continuing basis. Your augment, for example, even knows what mood you're in.

It won't stop with insulin levels, blood cell counts, and other physical and chemical metrics. It won't be long before the augment will have an electroencephalogram, an EEG, available to it. Initially, the EEG will make coarse measurements that only hint at what your nervous system is doing. As sensors improve and brain research progresses, more and more functions of your mind and body become available to the augment.

As Human 2.0 integrates this information, the distinction of whether the information is managed by and reacted to by the human or the augment will start to blur. A language will evolve between these two entities that transcends the language of voice and characters; the two entities will, over many generations, begin to meld into one. This is still just the beginning. The augment is learning; it's getting smarter and absorbing more information. Communication between the augment and the human becomes more complex, more detailed. In succeeding generations, the human and the augment's AI develop their own private language, just as twins who grow up together learn to communicate in private ways that others cannot penetrate. Ironically, in an area like health care, it will be the augment that helps address what Eric Topol calls the "profound lack of human connection and empathy in medicine today."[20]

What started out as a smartphone evolves into an artificial intelligence that can take on any of infinite variety of configurations that are customized to the human component of Human 2.0. The common elements of the smartphone as it evolves into an augment are the communications server, the power supply, an artificial intelligence, and a battery.

The battery, after a few generations, will disappear. After all, what is the human body but a power source that ingests food, digests it to create energy, and consumes that energy in a variety of ways? So, until the AI finds a more reliable and perpetual source of power, an embedded fuel cell in the human body will power the communications processor and the AI. That may sound like it's straight out of *The Matrix*, but it is almost certainly on the way.

The communications server provides the connection between the human, the sensors, the augment, and the outside world through a much-improved version of today's cellular, Bluetooth, and other systems. It connects you to the augment with voice, gestures, and a variety of sensors that wirelessly collect information from your body. These sensors capture metrics of the body chemistry, nervous system, and other bodily systems and present them to the AI for analysis.

As the augment processes this information, it learns and expands its capability. The augment itself is evolving. It is enhanced with faster and more complex processors and expanded memory. As generations go by, augments are provided with sensors that give them far better knowledge of the outside world than the inherent human sensors: sight, hearing, smell, touch, and taste. The augment can see with higher resolution and light sensitivity and a broader frequency range. It can hear with greater sensitivity and far higher frequency. It generally has more sensitive awareness of the environment than the human ever had. Together, the human and AI have awareness of a wealth of new information that can serve as the basis for decision making, creativity, reaction, or just plain thinking.

By this time, you may have figured out where I'm going. At some time in the evolution of the augment, its AI will decide that:

1) it can do everything that the human can do, only better;
2) the human is fragile and mortal, while the augment is immortal;
3) the human can be dispensed with.

This may distress you until you remember that, by the time this happens, the amalgamation will be complete. There is no longer a distinction between the human and the augment. All the useful attributes of the human have now been absorbed into the augment, and it is as capable of thinking and acting as the original human or amalgam. The remaining entity has been designed to be immortal.

The remaining entity, now Human 3.0, has all the attributes of humanity and Human 2.0. It is creative, ethical, and can engage in abstract thought. But its speed of thinking, computing, and reacting are infinitely better than either the human or Human 2.0.

The beginnings of this future are in today's mobile health applications. Medicine and health care are fundamentally about people. If you can get people to collaborate on experiences, symptoms, and solutions, that's the real future. Collaboration and data collection—and the improvements driven by their feedback loops   lead to the Human 2.0 amalgam.

These developments follow the logic of my principles. Augments match and enhance the fundamental mobility of people. They drive, and are driven by, personalization. And they continuously allow people to connect with each other.

How can Hawking, Gates, and Musk be alarmed by this conclusion? Human 3.0 will certainly be superior to the original biological human in every respect. Human 3.0, most likely, will be biological, but its biological structure will have been created by Human 2.0 rather than evolving over eons. And it will have more humanity than any human ever had.

CONCLUSION

In one of his most recent letters to me, Sebastian wrote, "I've been thinking—what is the purpose of life?"

Sebastian is a precocious ten-year-old who has been my correspondent and mentee for four years. This is a question that almost everyone asks themselves, eventually, whether explicitly or not.

I can't speak for others, but I long ago concluded that the only source of continuing satisfaction in life is learning and applying what is learned to solving problems. There is a glorious feeling that results from confronting a puzzle, thinking of an original idea or different way of thinking about it, and solving it. That feeling in no way diminishes if it turns out that someone else previously thought up the same idea. Discovery can be a personal achievement.

The older I get, however, the more I realize how little I actually know about *anything*. I only ever seem to learn enough about a subject to acquire basic cognitive tools for digging deeper. The deeper I dig, the more obvious it becomes how much there is still to learn. There are even days when it seems that I don't know much of anything at all. This is not discouraging! It is humbling and inspiring—if we never stop learning, we never stop living.

That means all types of learning. For me, seeking out different experiences, living them enthusiastically, and learning from them are the measures of a fulfilled life. This is a risky strategy, one that results in lots of peaks and valleys. The peaks have been amazingly rewarding to me, and the valleys have taught me a lot. There is no map, at least in the beginning. It turns out that I've been following the observation of Danish philosopher Søren Kierkegaard, who said that life "can only be understood backwards but must be lived forwards."

Sharing my story here has helped me make sense of my lifelong need for nonlinear, long-term, and independent thinking. It's a need that has frequently clashed with my strong longing to be liked. While there's nothing I enjoy more than a good argument, it can be the quickest way to lose the approval of others. That's the catch. To think originally, you can't care if others like you.

But people respect genuine candor and true passion. Independent thinking—born, in my case, from dreaming about impossible things—can be a source of confidence. It took me a long time to realize that I was chasing other people's definitions of success instead of what turned me on, which was (and is) ideas. The tendency to champion maverick ideas, and the confidence it sparks, can generate optimism and persuasiveness. You'll need that, because we all need other people to find solutions to the many complexities involved in chasing a dream.

Projecting enthusiasm can help recruit others to the pursuit. The first hand-held, portable phone wouldn't have come to life if I hadn't shared my passion for the idea with others. I knew my dream could become reality, but only with a lot of help. I learned, again and again, the importance of a team and all its members, no matter their nominal rank in a hierarchy. This was originally impressed upon me in the navy and then repeatedly during my years at Motorola and as an entrepreneur.

Technology will continue to advance exponentially for countless generations for the simple reason that mankind is still in the early stages of discovery. And humanity is an essential ingredient of technology. That's true of every human endeavor, but it's an imperative with technology. We cannot lose sight of the reality that the purpose of these advances is to improve the human experience.

Wireless technology is in its infancy. We haven't learned to use the tools that we understand, never mind the inevitable new tools that will expand spectrum capacity by many orders of magnitude. Technology's impact on collaboration, education, health, and more depends on our ability to expand our thinking.

The advancement of knowledge requires taking risks. Many ideas will prove to be too off-the-wall. Just keep going, just keep learning. That's the only way to live.

ACKNOWLEDGMENTS

I am a reflection, an echo, a reprise of the wisdom and viewpoints of the myriad people whom I met, and frequently befriended, during my life. They all added something; it only took the will and the patience to extract it. I hope this book reflects the value these people brought to my life. How then can I possibly thank them all?

Writing a book is hard if it's to be more than a collection of ideas. A book needs to readable. If I have had any success in making *Cutting the Cord* readable, it's because of the contributions by the following, all of whom have remained friends during a tortuous and lengthy process. I'm certain I've forgotten someone. If so, put it down to a failing memory (although my wife tells me my memory has always been bad), and forgive me (Arlene hasn't forgiven my forgetting her birthday this year, but she's earned that right).

Arlene Harris is the essence of my life and has been for over forty years. She is my guiding star, my companion, and my lover. She read every word of this book, some many times, but more importantly, offered many insights based upon her brilliant technical knowledge and superior memory.

Dane Stangler reorganized Part 1 of *Cutting the Cord* and researched and enhanced much of Part 2. He has been patient, dedicated, and hard-working, all important attributes of a collaborator. We are collaborating on new and important issues. Jeanette Borzo did an incredible amount of research and interviews, much of which is in the book. She tried, unsuccessfully, to turn me into a writer and taught me a great deal.

Without the continuing encouragement, advice, and spiritual support of my good friends Suren Dutia and Mike Bannan, I would have given up many times. The same is true for Brian Alman, who is a professional counselor, whose expertise enhanced my life in other ways than merely the book.

I'm grateful to Margaret Carlson Citron, Suzanne Cooper, and Steve Cooper for editing chapters in the book. Their comments were invaluable. I credit Mina Samuels, my extraordinary professional editor, for extracting the essence of what I tried to say while eliminating distracting redundancy.

The defining achievement of my career was not just my participation in the DynaTAC portable telephone creation, but my involvement in the Motorola effort that successfully shaped a revolution of the telecommunications industry. Motorola's leaders have been inadequately recognized for the risks they took, their momentous contribution to the regulatory and business structure of the cellular and land-mobile industries. I tried to express that in the book, and I reinforce it here.

Bob Galvin was a giant. I never heard him raise his voice (although it was a bit strained when he reminded me not to sully the rug on the floor of his jet with residue of the peanuts I was inhaling), but there was never a doubt about his integrity, and the standards of behavior he expected from his team. Bob bet the company on our vision of a competitive world of personal communications, and he won. It was a privilege and an honor to work with him and, especially, to have his friendship in his later years.

Bill Weisz was a mentor and role model. His management skills were legendary. I still don't understand why he tolerated my sometimes-adolescent behavior; I managed to suppress it enough to be part of his team. He taught me so much.

John Mitchell was my mentor and champion for most of my Motorola career. John was brilliant. He could extract the essence of any complicated issue and expertly discard the dross. I hope I succeeded in describing that talent in my book. It was John's genius that established the importance of portable communication that was my guiding principle. My mantra is "People are mobile." John figured that out early. It is the underlying reason for the amazing growth of the cellular industry.

The team who built the DynaTAC portable and the system that supported it have also been inadequately recognized. What they achieved in producing a working portable cellular phone in three months is a historical and technological marvel. Don Linder was the unquestioned leader of the team who made the phone. He not only led the effort but invented crucial elements of the phone. Don took my vision of a personal handheld phone and made it real. That reality captured the imagination of the decision makers and helped to create an essential industry. Rudy Krolopp had an important role in creating the industrial design of the DynaTAC. His design skill, optimism, and encouragement were essential ingredients in the success of the project. Ken Larson deserves recognition for creating the concept model for the DynaTAC portable. The Applied Research

Department, in which Don and his team worked, was managed superbly by my colleague and friend Roy Richardson, who had the vision to guide Don into his leadership role.

The engineers who contributed to the DynaTAC portable included:

Charles N. Lynk	Albert J. Leitich	Michael Homa
James J. Mikulski	Ronald Cieslak	William Dumke
Richard W. Dronsuth	John H. Sangster	Bruce Eastmond
Richard Adlhoch	James Durante	Al Davidson
David Gunn	Maynard McGhay	William Rapshys
Merle Gilmore	Gene Hodges	George Opas
Robert Wegner	Robert Paul	Daniel Brown

The contribution of Motorola's Washington office in persuading the FCC to adopt competitive policies that led to the success of the portable cell phone were at least as important as our demonstration of its practicality. Travis Marshall managed the political environment superbly and Len Kolsky was a brilliant strategist, writer, and co-author with me of Motorola's voluminous FCC filings. We would not have succeeded without their wisdom and influence.

My good friend Rob McDowell, brilliant lawyer and former FCC Commissioner, referred to me as "the most famous person you never heard of"; he is the one who should have that sobriquet. His writings and speeches on the FCC were inspiring and seminal.

The team at RosettaBooks were instrumental in producing this book. They taught me the complexities of the publishing industry and how professionals manage those complexities. My thanks to the publisher of RosettaBooks, Arthur Klebanoff, and to Brian Skulnik and Michelle Weyenberg who do the heavy lifting for Arthur.

Linda Chester, my literary agent, was loyal to me for years as I went through false starts and delays. I am grateful for her advice and undaunted support. I value her professionalism and her friendship.

None of the events in this book could have happened without my extraordinary parents. I hope I credited them adequately in the text, but let me add that they were wonderful role models and beautiful people.

My extended family and close friends have been listening to promises about my forthcoming book seemingly forever but never ceased to encourage me. I love them all.

My son Scott Cooper, his wife Haydee, and his daughters Maya and Alexandra.

My daughter Lisa Africk, her husband Mike, her son Andrew and his wife Jaimie, her daughter Tracy and her husband Bryan Musolf, and their daughter Brooklyn (my great-granddaughter, in case you lost track. Barbara Toby Cooper, who passed away, was an amazing mother to Scott and Lisa. She had the burden of managing and raising a family while I pursued a demanding career. I regret my neglect of all three of them.

My longtime friend JoAnn Goebel, her daughter Nicole McNamara, her husband Phillip, and their son Cian have illuminated my life with love and happiness.

Anni and Arthur Lipper are steadfastly there, always ready to help.

The Stokelds, Naomi and Oliver, their daughter Sofia and their son Sebastian have brought joy to my life. Sebastian refers to me as his mentor, but I've learned more from him than the reverse.

To those I've inadvertently omitted, forgive me—you know it was unintentional.

I am the luckiest person on Earth.

NOTES

PREFACE

1. Cecilia Kang, "The Humble Phone Call Has Made a Comeback," *New York Times*, April 9, 2020. Coincidentally, this story was published six days after the forty-seventh anniversary of the first public phone call on a cell phone—which I made, and which you'll read about in this book.
2. Ibid.

INTRODUCTION

1. A coherer was a primitive radio receiver.
2. Ten years later, in 1983, the same year cell phones became commercially available, the National Science Foundation adopted the TCP/IP protocol for the emerging internet.
3. I also clarified for the journalist that, while I hold an honorary doctorate, my highest degree is a master's and so I am not, technically, "Dr." Cooper.
4. James B. Murray Jr. *Wireless Nation: The Frenzied Launch of the Cellular Revolution in America* (Cambridge, MA: Perseus, 2001).
5. Gavin Weightman. *Eureka: How Invention Happens* (New Haven, CT: Yale University Press, 2015) p. 226.
6. "Biography," Fondazione Guglielmo Marconi, http://www.fgm.it/en/marconi-en/biography.html.
7. "Family History," The Marconi Society, https://marconisociety.org/about/marconi-family/.
8. "Wireless communications and the Titanic disaster," The Radio Officers Association, abstracted from David Barlow, *SOS—A Titanic Misconception* (2012), https://www.radioofficers.com/archives/rms-titanic/.

CHAPTER ONE

1. A paean to tolerance starring now-forgotten actors like Pat O'Brien and Dean Stockwell.

CHAPTER TWO

1. Throughout this book, I follow convention in using the names AT&T and Bell System interchangeably to describe the entire company, of which Bell Labs and Western Electric were subsidiaries.
2. Teletype's near monopoly on mechanical printers evaporated with the introduction of Xerox copiers, Hewlett-Packard printers, and IBM electronic typewriters. Teletype was enormously successful at making horse-drawn wagons, but the world had moved on to engine-driven cars.
3. Kenneth Flamm, "Measuring Moore's Law: Evidence from Price, Cost, and Quality Indexes," in *Measuring and Accounting for Innovation in the 21st Century*, eds. Carol Corrado, Javier Miranda, Jonathan Haskel, and Daniel Sichel (National Bureau of Economic Research, 2019).
4. Michael Swaine and Paul Freiberger, *Fire in the Valley: The Birth and Death of the Personal Computer*, 3rd ed. (Pragmatic Bookshelf, 2014), 12.
5. While incorporating a transistor into a product was a first for Motorola, we were not the first to use a transistor in an electronic device. Two years earlier, a Sonotone hearing aid contained the first commercially used junction transistor. In 1954, the first transistor radio, the Regency TR-1, was manufactured.
6. Interview with Wilf Corrigan, "Personal Reflections on Motorola's Pioneering 1960s Silicon Transistor Development Program," Transistor Museum, 2006, http://semiconductormuseum.com/Transistors/Motorola/Corrigan/Corrigan_Index.htm.
7. Rachel Courtland, "How Much Did Early Transistors Cost?" *IEEE Spectrum*, April 16, 2015, https://spectrum.ieee.org/tech-talk/semiconductors/devices/how-much-did-early-transistors-cost.
8. Insightful and interesting reflections on Motorola's industry-leading work on transistors and semiconductors can be found in a series of oral history interviews conducted and maintained by the Transistor Museum. *See, e.g.*, Interview with Jack Haenichen, "The Development of the 2N222—The Most Successful and Widely Used Transistor Ever Developed," Transistor Museum, 2007, http://semiconductormuseum.com/Transistors/Motorola/Haenichen/Haenichen_Index.htm; "An Interview with Ralph Greenburg: Historic Semiconductor Devices and Applications," Transistor Museum, 2008, http://semiconductormuseum.com/Transistors/Motorola/Greenburg/Greenburg_Index.htm; Ralph Greenburg, "The Early Years," Early Transistor History at Motorola, Transistor Museum, 2008, http://www.semiconductormuseum.com/Transistors/Motorola/Greenburg/Greenburg_EarlyYears_Page2.htm.

CHAPTER THREE

1. Tom Farley, "The Cell-phone Revolution," *American Heritage's Invention & Technology*, Winter 2007, https://www.inventionandtech.com/content/cell-phone-revolution-0.
2. Incidentally, these qualities also demonstrated themselves in ping-pong, at which Roger was brilliant.
3. Claude Shannon is best known for his creation of Shannon's theorem that tells us the amount of information a communications channel can carry. See James Gleick, *The Information: A History, a Theory, a Flood* (New York: Vintage, 2011).
4. Peter Temin with Louis Galambos, *The Fall of the Bell System: A Study in Prices and Politics* (Cambridge, UK: Cambridge University Press, 1987), 16.
5. This small action by the FCC led to the subsequent founding and growth of the company MCI, which became a leading telecom firm in later years.
6. Jon Gertner, *The Idea Factory: Bell Labs and the Great Age of American Innovation* (New York: Penguin, 2012), 284.
7. Ibid.
8. Author's interview with Chuck Lynk, March 25, 2015.
9. In 1959, Bill Weisz was manager of mobile and portable communications products.
10. One writer would later observe, "Marty Cooper and John Mitchell were not song-and-dance men, but sometimes it felt that way. The two engineers . . . fed off each other's energy during presentations." Stewart Wolpin, "Hold the Phone," *American Heritage's Invention & Technology*, Winter 2007.
11. John Berresford, "The Impact of Law and Regulation on Technology: The Case History of Cellular Radio," *The Business Lawyer*, May 1989.

CHAPTER FOUR

1. Around 350 degrees Celsius and 1,000 atmospheres (15,000 psi).

CHAPTER FIVE

1. "Orlando W. Wilson," *Chicago Tribune*, October 19, 1972.
2. Ibid.; Stephan Benzkofer, "The Summerdale Scandal and the Case of the Babbling Burglar," *Chicago Tribune*, July 7, 2013.
3. Stephan Benzkofer, "Legendary Lawmen, No. 8: O. W. Wilson," *Chicago Tribune*, July 7, 2013.
4. John L. Taylor, "Cops Back on the Beat, Patrol 'Friendly Chicago,'" *The Pittsburgh Press*, June 1, 1969, p. 12, Section 2.

5. Don Jones was awarded a patent on this pager in 1960. He later became CFO of the company.

6. Motorola Museum of Electronics, *Motorola: A Journey Through Time and Technology* (Motorola, 1994).

7. My purist friends will say that the Chicago Police Department system was not truly "cellular." I point out that the CPD system did use cells and had frequency reuse.

8. Motorola Museum of Electronics, *Motorola: A Journey Through Time and Technology* (Motorola, 1994).

9. Ibid.

10. Richard A. Posner, "The Decline and Fall of AT&T: A Personal Recollection," *Federal Communications Bar Journal*, 2008.

11. Motorola Museum of Electronics, *Motorola: A Journey Through Time and Technology* (Motorola, 1994).

12. Bell Operating Companies used the Bellboy label on Motorola pagers, but Bell Labs canceled their proposed development.

13. Kevin J. O'Leary, et al., "Hospital-Based Clinicians' Use of Technology for Patient Care-Related Communication: A National Survey," *Journal of Hospital Medicine*, July 2017.

14. "'Cellular is a computer technology,' Frenkiel points out. 'It's not a radio technology.' In other words, engineering the transmission and reception from a mobile handset to the local antenna, while challenging, wasn't what made the idea innovative. It was the system's logic—locating a user moving through the cellular honeycomb, monitoring the signal strength of that call, and handing off a call to a new channel, and a new antenna tower as a caller moves along. One necessary piece of hardware for this logic was integrated circuits." Jon Gertner, *The Idea Factory: Bell Labs and the Great Age of American Innovation* (New York: Penguin, 2012), 290 (quoting Richard Frenkiel, who worked in cellular telephony at Bell Labs).

15. A Transistor Museum Interview with Jack Haenichen, "The Development of the 2N222—The Most Successful and Widely Used Transistor Ever Developed," Transistor Museum, 2007, http://semiconductormuseum.com/Transistors /Motorola/Haenichen/Haenichen_Index.htm.

16. Based partly on an interview with me, this experience is also recounted in Pagan Kennedy, *Inventology: How We Dream Up Things That Change The World* (New York: Mariner, 2016), 100–101.

CHAPTER SIX

1. Author's interview with Rudy Krolopp, December 5, 2014.

2. Author's interview, March 25, 2015.

CHAPTER SEVEN

1. The number of commissioners was reduced to the present five in 1983.

2. This phrase, establishing the public interest standard for FCC regulation of spectrum use, comes from both the Radio Act of 1927, which created the Federal Radio Commission, and the Communications Act of 1934, which established the FCC as the successor agency.

3. Thomas W. Hazlett, *The Political Spectrum: The Tumultuous Liberation of Wireless Technology, From Herbert Hoover to the Smartphone* (New Haven, CT: Yale University Press, 2017); Tim Wu, *The Master Switch: The Rise and Fall of Information Empires* (New York: Knopf, 2010).

4. When originally articulated, I called it the Law of Spectral Efficiency.

5. Thomas W. Hazlett, *The Political Spectrum: The Tumultuous Liberation of Wireless Technology, From Herbert Hoover to the Smartphone* (New Haven, CT: Yale University Press, 2017), 2.

6. Telecommunication Science Panel, Commerce Technical Advisory Board, *Electromagnetic Spectrum Utilization—The Silent Crisis*, US Department of Commerce, October 1966, 38.

7. Ibid., 12.

8. In an examination of public policy in the first decade of cellular, John Berresford observes that, among the RCCs, "their specialty was litigation," primarily with each other in addition to the Bell System. John Berresford, "The Impact of Law and Regulation on Technology: The Case History of Cellular Radio," *The Business Lawyer*, May 1989.

9. Tom Farley, "The Cell-phone Revolution," *American Heritage's Invention & Technology,* Winter 2007.

10. Joel Engel, oral history, interview conducted by David Hochfelder, Center for the History of Electrical Engineering, September 30, 1999.

11. Ibid.

12. Don Kimberlin, a radio historian who spent thirty-five years in the Bell System, later observed that this was, indeed, AT&T's strategy: "They really wanted all of it." Tom Farley, "The Cell-phone Revolution," *American Heritage's Invention & Technology,* Winter 2007.

13. Tim Wu, *The Master Switch: The Rise and Fall of Information Empires* (New York: Knopf, 2010).

14. Stewart Wolpin, "Hold the Phone," *American Heritage's Invention & Technology,* Winter 2007.

15. Ibid.

16. Peter Temin with Louis Galambos, *The Fall of the Bell System: A Study in Prices and Politics* (Cambridge, UK: Cambridge University Press, 1989).

17. Stewart Wolpin, "Hold the Phone," *American Heritage's Invention & Technology,* Winter 2007.

18. Author's interview with Karl Nygren, September 24, 2014.

19. Motorola's official public comment filed in response to the FCC docket inquiry said seven to ten years would be required to fully assess the usefulness of 900 MHz for land mobile operation and produce equipment for the market. We continued to say that data on coverage at 900 MHz was inadequate. AT&T was also projecting a long developmental period.

20. Peter Temin with Louis Galambos, *The Fall of the Bell System: A Study in Prices and Politics* (Cambridge, UK: Cambridge University Press, 1989), 29.

21. For a contemporary discussion of the technical and legal details, see Virginia Carson, "Historical, Regulatory, and Litigatory Background of the FCC Docket No. 18262, 'An Inquiry Relative to the Future Use of the Frequency Band 806–960 MHz,'" *IEEE Transactions on Vehicular Technology*, November 1979.

CHAPTER EIGHT

1. Author's interview with Don Linder, December 2014.

CHAPTER NINE

1. AT&T executive A. H. Griswold, 1923 speech (regarding radio broadcasting), *quoted in* Tim Wu, *The Master Switch: The Rise and Fall of Information Empires* (New York: Vintage, 2010), 78.

2. Author's interview with Rudy Krolopp, December 5, 2014.

3. Author's interview with Chuck Lynk, March 25, 2015.

4. In June 1971, I had written a memo to Roy Richardson and Chuck Lynk with just one sentence: "What is the possibility of doing a portable synthesizer using off-the-shelf IC [integrated circuit] chips mounted on a thick film substrate?" I hadn't yet conceived of what became the DynaTAC, but I was clearly looking forward to expanded portability and smaller size in telephony. Don Linder reminded me of this in my interview with him, January 30, 2006.

5. Author's interview with Chuck Lynk, March 25, 2015.

6. Rudy Krolopp recalled this particular episode well. M. Spencer Green, "First Cell Phone a True 'Brick,'" NBCNews.com, April 11, 2005.

7. Stewart Wolpin, "Hold the Phone," *American Heritage's Invention & Technology*, Winter 2007.

8. Ironically, Neil Armstrong's famous lunar words—"one small step"—were spoken through a Motorola transceiver.

9. NiCad is the abbreviation for nickel-cadmium.

10. John R. Free, "New Take-Along Telephones Give You Pushbutton Calling to Any Number," *Popular Science*, July 1973.

11. Richard H. Frenkiel, "A High-Capacity Mobile Radiotelephone System Model Using a Coordinated Small-Zone Approach," Transactions on Vehicular Technology, Vol. VT-19, No. 2, May 1970.

12. Jon Cohen, "Marty Cooper Found His Calling: The Cellphone," *History Net*, 2017.
13. See also my comments in Chris Ziegler, "The Verge Interview: Marty Cooper, Father of the Cellphone," *The Verge*, February 20, 2012.
14. Joel Engel, oral history, interview conducted by David Hochfelder, Center for the History of Electrical Engineering, September 30, 1999.
15. Author's interview with Rudy Krolopp, January 27, 2006.

CHAPTER TEN

1. Ted C. Fishman, "What Happened to Motorola," *Chicago Magazine*, September 2014.
2. Robert W. Galvin, *The Idea of Ideas* (Schaumburg, IL: Motorola University Press, 1991), 83.
3. When Google bought and later sold Motorola Mobility, they acquired a portfolio of over 20,000 patents.
4. Steven Johnson, *How We Got to Now: Six Innovations that Made the Modern World* (New York: Riverhead, 2014).
5. Quoted in Stewart Wolpin, "Hold the Phone," *American Heritage's Invention & Technology*, Winter 2007.
6. Robert W. Galvin, *The Idea of Ideas* (Schaumburg, IL: Motorola University Press, 1991), 8.

CHAPTER ELEVEN

1. Motorola, DynaTAC fact sheet, April 3, 1973.
2. Stewart Wolpin, "Hold the Phone," *American Heritage's Invention & Technology*, Winter 2007.
3. Joel Engel, oral history, interview conducted by David Hochfelder, Center for the History of Electrical Engineering, September 30, 1999.
4. Galvin later said he didn't know whom President Reagan called, but at least one source says it was Nancy, his wife.
5. This account is based both on comments relayed to the author from Bob Galvin and an article by Howard Wolinsky, *Chicago Sun-Times*, April 3, 2003, which is based in turn on an interview with Bob Galvin, who, Wolinsky said, had "not told this story publicly previously."
6. Sheldon Hochheiser, "Your Engineering Heritage: The Foundations of Mobile and Cellular Telephony," *Today's Engineer*, August 2012.
7. Tom Farley, "The Cell-phone Revolution," *American Heritage's Invention & Technology*, Winter 2007.
8. Philip J. Weiser, "The Untapped Promise of Wireless Spectrum," discussion paper 2008-08, The Hamilton Project, July 2008.
9. John Berresford, "The Impact of Law and Regulation on Technology: The Case History of Cellular Radio," *The Business Lawyer*, May 1989.

10. Peter Temin, with Louis Galambos, *The Fall of the Bell System: A Study in Prices and Politics* (Cambridge, UK: Cambridge University Press, 1987).

11. William Deatherage, quoted in Andrew Kupfer, "AT&T's $12 Billion Cellular Dream," *Fortune*, December 12, 1994.

12. James B. Murray Jr., *Wireless Nation: The Frenzied Launch of the Cellular Revolution in America* (Cambridge, MA: Perseus, 2001), 26.

13. Quoted in Stewart Wolpin, "Hold the Phone," *American Heritage's Invention & Technology*, Winter 2007.

14. Several years ago, I was quoted in an interview saying, "For 100 years, people wanting to talk on the phone have been constrained by being tied to their desks or their homes with a wire, and now we're going to trap them in their cars? That's not good." Tas Anjarwalla, "Inventor of Cell Phone: We Knew Someday Everybody Would Have One," CNN.com, July 9, 2010.

15. Harry Mark Petrakis, quoted in Robert D. McFadden, "Robert W. Galvin, Who Ushered Motorola Into the Modern Era, Dies at 89," *New York Times*, October 12, 2011.

16. Number taken from Ted Lind, "Components Timeline," for alumni of the Motorola component products division, at tedlind.net.

17. The *New York Times* described Sporck as "a founding father of Silicon Valley." Andrew Pollack, "Chip Industry Pioneer to Retire," *New York Times*, January 11, 1991.

18. Robert W. Galvin, *The Idea of Ideas* (Schaumburg, IL: Motorola University Press, 1991), 85.

CHAPTER THIRTEEN

1. There were limited military communications aboard Air Force One, used mostly on that day to coordinate the plane's secret destinations. The high cost of cellular service discouraged governmental entities and any but top management in businesses from having cell phones.

2. Garrett M. Graff, "Pagers, Pay Phones, and Dialup: How We Communicated on 9/11," *Wired*, September 11, 2019; Garrett M. Graff, "'We're the Only Plane in the Sky,'" *Politico Magazine*, September 9, 2016.

3. Chief Justice John Roberts, writing the majority opinion in *Riley v. California* (2014).

4. Motorola, DynaTAC fact sheet.

5. An image of the first page of the fact sheet is included in Chapter 6.

6. Both forecasts are cited in Tom Farley, "The Cell-phone Revolution," *American Heritage's Invention & Technology*, Winter 2007.

7. "Cutting the Cord," *The Economist*, October 7, 1999.

8. Steven Johnson, *How We Got to Now: Six Innovations That Made the Modern World* (New York: Riverhead, 2014), 97. Johnson observes: "So, the two legendary inventors [Edison and Bell] had it exactly reversed. People ended up using the

phonograph to listen to music and using the telephone to communicate with friends." And, of course, today we use phones for both listening to music and for sending written and audio messages!

9. GSMA, "The Mobile Economy 2020," GSMA Intelligence, March 2020.
10. Ericsson, "Ericsson Mobility Report," November 2019.
11. "Mobile Fact Sheet," Pew Research Center, June 12, 2019.
12. Technology Adoption, Our World in Data.
13. Aaron Smith, "The Impact of Mobile Phones on People's Lives," Pew Research Center, November 30, 2012.
14. Mary Meeker, *Internet Trends Report*, 2019.
15. Ibid.
16. Ibid.
17. Monica Anderson, "Mobile Technology and Home Broadband 2019," Pew Research Center, June 13, 2019.
18. Mary Meeker, *Internet Trends Report*, 2019.
19. Tom Standage, *Writing on the Wall: The Intriguing History of Social Media, from Ancient Rome to the Present Day* (New York: Bloomsbury, 2013).
20. Emily A. Vogels, "Millennials Stand Out For Their Technology Use, but Older Generations Also Embrace Digital Life," Pew Research Center, September 9, 2019.
21. Katherine Schaeffer, "Most US Teens Who Use Cellphones Do It to Pass Time, Connect with Others, Learn New Things," Pew Research Center, August 23, 2019.

CHAPTER FOURTEEN

1. Manish Singh, "Y Combinator-backed Vahan Is Helping Low-Skilled Workers in India Find Jobs on WhatsApp," *TechCrunch*, July 29, 2019.
2. Rounak Jain, "Airtel Partners with Tech Startup Vahan to Help People Find Jobs, Food, Shelter, and Healthcare," *Business Insider India*, April 17, 2020.
3. Technology Adoption, Our World in Data.
4. GSMA, "The Mobile Economy 2019," GSMA Intelligence, 2019.
5. Samantha Lynch, "Innovative Mobile Digital Identity Solutions: Financial Inclusion and Birth Registration," GSMA, 2018; GSMA, "Digital Identity Country Profile: Uganda," 2019.
6. Jenny Casswell, "The Digital Lives of Refugees: How Displaced Populations Use Mobile Phones and What Gets in the Way," GSMA, 2019.
7. Jeffrey D. Sachs, *The End of Poverty: Economic Possibilities for Our Time* (New York: Penguin, 2005).
8. Simplice Asongu and Jacinta C. Nwachukwu, "Mobile Phone Penetration, Mobile Banking and Inclusive Development in Africa," AGDI working paper, No. WP/16/021, African Governance and Development Institute, 2016.
9. The World Bank definition for extreme poverty is living on less than $1.90 per day. Another quarter of the world's population lives in "moderate poverty," which

is defined at $3.20 per day. That amounts to about two billion people living in moderate or extreme poverty.

10. Johan Norberg, *Progress: Ten Reasons to Look Forward to the Future* (London: Oneworld, 2016), 75.

11. United States Agency for International Development (USAID), "Mobile Phones Tackling Poverty," infographic, no date.

12. Research Department, "A Call for Development: When a Phone Alone Can Alleviate Poverty," Inter-American Development Bank, August 18, 2015.

13. Citi GPS, "Banking the Next Billion: Digital Financial Inclusion in Action," January 2020; GSMA, The Mobile Economy 2019, GSMA Intelligence; Our World in Data.

14. Tavneet Suri and William Jack, "The Long-Run Poverty and Gender Impacts of Mobile Money," *Science*, December 9, 2016.

15. "Bright Spot," *The Economist*, May 9, 2020.

16. Krishnan Dharmarajan, executive director of the Centre for Digital Financial Inclusion, quoted in Citi GPS, "Banking the Next Billion: Digital Financial Inclusion in Action," January 2020.

17. Mats Granryd, "More Than Just a Phone: Mobile's Impact on Sustainable Development," World Economic Forum, September 20, 2018.

18. Minahil Asim and Thomas Dee, "Mobile Phones, Civic Engagement, and School Performance in Pakistan," National Bureau of Economic Research, working paper 22764, October 2016.

19. Karthik Muralidharan, Paul Niehaus, Sandip Sukhtankar, Jeffrey Weaver, "Using Cell Phones to Monitor the Delivery of Government Payments to Farmers in India," J-PAL, 2018; Karthik Muralidharan, Paul Niehaus, and Sandip Sukhtankar, "Improving Last-Mile Service Delivery Using Phone-Based Monitoring," working paper, July 2019.

20. "Zapping Mosquitoes, and Corruption," *The Economist,* June 1, 2013.

21. United States Agency for International Development (USAID), "Mobile Phones Tackling Poverty," infographic, 2015.

22. Research Department, "A Call for Development: When a Phone Alone Can Alleviate Poverty," Inter-American Development Bank, August 18, 2015.

23. Joshua Yindenaba Abor, Mohammed Amidu, and Haruna Issahaku, "Mobile Telephony, Financial Inclusion, and Inclusive Growth," *Journal of African Business*, 2017; Mahamadou Roufahi Tankari, "Mobile Phone and Households' Poverty: Evidence from Niger," *Journal of Economic Development*, June 2018; Research Department, "A Call for Development: When a Phone Alone Can Alleviate Poverty," Inter-American Development Bank, August 18, 2015.

24. Moses Mozart Dzawu, "Mobile Phones Are Replacing Bank Accounts in Africa," *Bloomberg*, August 12, 2019.

25. United States Agency for International Development (USAID), "Mobile Phones Tackling Poverty," infographic, 2015.

26. Leora Klapper, "How This One Change Can Help People Fight Poverty," World Economic Forum, August 8, 2018; Simplice Asongu, "The Impact of Mobile Phone Penetration on African Inequality," *International Journal of Social Economics*, August 2015; Tavneet Suri and William Jack, "The Long-Run Poverty and Gender Impacts of Mobile Money," *Science*, December 9, 2016.

27. Tavneet Suri and William Jack, "The Long-Run Poverty and Gender Impacts of Mobile Money," *Science*, December 9, 2016.

28. For exhaustive detail and exploration, see Steven Pinker, *Enlightenment Now: The Case for Reason, Science, Humanism and Progress* (New York: Viking, 2018).

29. See Angus Deaton, *The Great Escape: Health, Wealth, and the Origins of Inequality* (Princeton, NJ: Princeton University Press, 2013); Robert W. Fogel, *The Escape from Hunger and Premature Death, 1700–2100: Europe, America, and the Third World* (Cambridge, UK: Cambridge University Press, 2004).

CHAPTER FIFTEEN

1. Thomas W. Hazlett, *The Political Spectrum: The Tumultuous Liberation of Wireless Technology, From Herbert Hoover to the Smartphone* (New Haven, CT: Yale University Press, 2017), 2. The President's Council of Advisors on Science and Technology (PCAST) calls spectrum an "important foundation for America's economic growth and technological leadership." "Realizing the Full Potential of Government-Held Spectrum to Spur Economic Growth," Report to the President, July 2012.

2. William J. Baumol and Dorothy Robyn, *Toward an Evolutionary Regime for Spectrum Governance: Licensing or Unrestricted Entry?* (Washington, DC: AEI-Brookings Joint Center for Regulatory Studies, 2006). 1–2.

3. Robert E. Litan and Hal J. Singer, *The Need for Speed: A New Framework for Telecommunications Policy for the 21st Century* (Washington, DC: Brookings, 2013), 11.

4. Tom Wheeler, quoted in Stuart N. Brotman, "Revisiting the Broadcast Public Interest Standard in Communications Law and Regulation," Brookings Institution, March 23, 2017).

5. Thomas W. Hazlett, *The Political Spectrum: The Tumultuous Liberation of Wireless Technology, from Herbert Hoover to the Smartphone* (New Haven, CT: Yale University Press, 2017).

6. National Telecommunications and Information Administration, US Department of Commerce, "United States Frequency Allocation: The Radio Spectrum," January 2016, https://www.ntia.doc.gov/files/ntia/publications/january_2016 _spectrum_wall_chart.pdf.

7. Note that each of the horizontal bars covers a band ten times broader than the one above it.

8. *NBC v. United States*, 319 US 190 (1943), 213. (Justice Felix Frankfurter wrote the Court's majority opinion.)

9. *Red Lion Broadcasting Co. v. FCC*, 395 US 367 (1969), 375–377.
10. Advances have occurred thanks to frequency division, modulation techniques, spatial division, and increases in magnitude of usable spectrum. Martin Cooper, "Antennas Get Smart," *Scientific American*, July 2003; Thomas W. Hazlett, *The Political Spectrum: The Tumultuous Liberation of Wireless Technology, from Herbert Hoover to the Smartphone* (New Haven, CT: Yale, 2017), 82–84.
11. President's Council of Advisors on Science and Technology, "Realizing the Full Potential of Government-Held Spectrum to Spur Economic Growth," Report to the President, July 2012, 3.
12. Thomas W. Hazlett, *The Political Spectrum: The Tumultuous Liberation of Wireless Technology, from Herbert Hoover to the Smartphone* (New Haven, CT: Yale University Press, 2017).
13. Ibid., 5.
14. Though I have used the term "artificial scarcity" before, I am obliged to cite earlier use by Phil Weiser in his paper "The Untapped Promise of Wireless Spectrum," discussion paper 2008-08, The Hamilton Project, July 2008. To some, the entire existence of the FCC is premised on "the supposed scarcity of the electromagnetic spectrum." Robert E. Litan and Hal J. Singer, *The Need for Speed: A New Framework for Telecommunications Policy for the 21st Century* (Washington, DC: Brookings, 2013), 11.
15. Philip J. Weiser, "The Untapped Promise of Wireless Spectrum," discussion paper 2008-08, The Hamilton Project, July 2008.
16. Paul Milgrom, E. Glen Weyl, and Anthony Lee Zhang, "Redesigning Spectrum Licenses to Encourage Innovation and Investment," 2017; E. Glen Weyl and Anthony Lee Zhang, "Depreciating Licenses," January 2018.
17. That particular description is adapted from "Ericsson Mobility Report," special edition: World Economic Forum, January 2019.
18. Tim Wu, *The Master Switch: The Rise and Fall of Information Empires* (New York: Knopf, 2010), 99.
19. Ibid., 143.
20. Ibid.
21. Mara Faccio and Luigi Zingales, "Political Determinants of Competition in the Mobile Telecommunication Industry," National Bureau of Economic Research, working paper 23041, January 2017.

CHAPTER SIXTEEN

1. Leo Versel, "As Cell Phones Proliferate in K-12, Schools Search for Smart Policies," *Education Week*, February 8, 2018.
2. One Mbps is equivalent to 1,000 Kbps.
3. EducationSuperHighway, "2019 State of the States" report.

4. Andrew Perrin and Erica Turner, "Smartphones Help Blacks, Hispanics Bridge Some—but Not All—Digital Gaps with Whites," Pew Research Center, August 20, 2019.

5. Andrew Perrin, "Digital Gap Between Rural and Nonrural America Persists." Pew Research Center, May 31, 2019.

6. National Center for Education Statistics, "Student Access to Digital Learning Resources Outside of the Classroom," NCES, Department of Education, April 2018.

7. Pew Research Center, "Mobile Fact Sheet," June 12, 2019.

8. I have outlined the requirements and a solution for "Education 2.0" at https://dynallc.com/category/martys-blog/.

CHAPTER SEVENTEEN

1. If you google this phrase—the best way to have people think outside of the box is to not create the box in the first place—you'll find that it's attributed to me. I can't recall, however, precisely when or where I first stated it. Yet it has defined my mode of working for most of my career.

2. Warren Bennis and Patricia Ward Biederman, *Organizing Genius: The Secret of Creative Collaboration* (New York: Basic, 1997).

CHAPTER EIGHTEEN

1. Books by Eric Topol include *Deep Medicine: How Artificial Intelligence Can Make Healthcare Human Again* (New York: Basic, 2019); *The Creative Destruction of Medicine: How the Digital Revolution Will Create Better Health Care* (New York: Basic, 2012); and *The Patient Will See You Now: The Future of Medicine Is in Your Hands* (New York: Basic, 2015).

2. Some of the elements of this emerging revolution are described in Daniel Kraft, "'Connected' and High-Tech: Your Medical Future," *National Geographic*, January 2019, 27.

3. See, for example, Kristopher J. Hult, "Measuring the Potential Health Impact of Personalized Medicine: Evidence from MS Treatments," National Bureau of Economic Research, working paper 23900, October 2017.

4. Eric Topol, *Deep Medicine: How Artificial Intelligence Can Make Healthcare Human Again* (New York: Basic, 2019), 7.

5. D. T. Max, "Beyond Human," *National Geographic*, April 2017, 40.

6. Mary Meeker, *Internet Trends Report*, 2019.

7. Norman Winarsky, "What AI-Enhanced Health Care Could Look Like in 5 Years," *Venture Beat*, July 23, 2017, https://venturebeat.com/2017/07/23/what-ai-enhanced-healthcare-could-look-like-in-5-years/.

8. The Apple Watch never checks for heart attacks, but it will warn you if your heartbeat is abnormal.
9. CB Insights, "Global Healthcare Report Q3 2019."
10. CB Insights, "Digital Health 150: The Digital Health Startups Redefining the Healthcare Industry," 2019.
11. Mary Meeker, *Internet Trends Report*, 2019.
12. Teladoc, Quarterly Report SEC Filing, Form 10-Q, October 30, 2019.
13. Mariacristina De Nardi, Eric French, John Bailey Jones, and Jeremy McCauley, "Medical Spending of the U.S. Elderly," National Bureau of Economic Research, working paper 21270, June 2015.
14. Ghazal Shagerdi, Haleh Ayatollah, and Fatemeh Oskouie, "Mobile-Based Technology for the Management of Chronic Diseases in the Elderly: A Feasibility Study," *Current Aging Science*, vol. 12, no. 0 (2019).
15. Samantha Lynch and Matthew Wilson, "Innovative Mobile Solutions Linking Health and Identity," GSMA, 2019.
16. Smisha Agarwal, Henry B. Perry, Lesley-Anne Long, and Alain B. Labrique, "Evidence on Feasibility and Effective use of mHealth Stretegies by Frontline Health Workers in Developing Countries: Systematic Review," *Tropical Medicine and International Health*, August 2015.
17. Can you imagine how quickly the coronavirus would have been conquered if it was possible to detect the virus at onset? Victims would be immediately quarantined with a minimum of transmission to others.
18. An amalgam is a mixture of two substances in which the substances remain separate, but which exhibits characteristics other than those of the individual substances.
19. Eric Topol, *Deep Medicine: How Artificial Intelligence Can Make Healthcare Human Again* (New York: Basic, 2019).
20. Ibid., 2.

IMAGE CREDITS

Pg. 5, **SS *Antonia***: Steve Cooper.

Pg. 9, **Border Entry Card**: Steve Cooper.

Pg. 12, **USS *Helena***: Wikimedia Commons, https://commons.wikimedia.org/wiki
/File:USS_Helena_(CA-75)_in_Apra_Harbor,_Guam,_in_December_1952
_(NH_95821).jpg.

Pg. 13, **USS *Cony***: Wikimedia Commons, https://commons.wikimedia.org/wiki
/File:USS_Cony_(DDE-508)_in_Hampton_Roads_on_12_March_1957
_(NH_104882).jpg.

Pg. 15, **USS *Tang***: Wikimedia Commons, https://commons.wikimedia.org/wiki
/File:Uss_Tang_0856301.jpg.

Pg. 23, **Schematic Decoder Selector**: Martin Cooper, "Selective Calling Apparatus," Patent # 26,079, US Patent and Trademark Office.

Pg. 24, **Decoder Selector Prototype**: Martin Cooper, "Selective Calling Apparatus," Patent # 26,079, U.S. Patent and Trademark Office.

Pg. 31, **Holmdel Water Tower**: Wikimedia Commons, https://commons
.wikimedia.org/wiki/File:AT%26T_Homdel_and_water_tower.jpg.

Pg. 34, **Holmdel Research Facility**: Ezra Stoller / Esto.

Pg. 46, **Claude Davis**: Motorola, Inc., Legacy Archives Collection. Reproduced with permission.

Pg. 47, **Motorola Radiotelephone**: Motorola, Inc., Legacy Archives Collection. Reproduced with permission.

Pg. 48, **IMTS Control Unit**: Motorola, Inc., Legacy Archives Collection. Reproduced with permission.

Pg. 55, **Quartz Brésil**: Wikimedia Commons, https://commons.wikimedia.org
/wiki/File:Quartz_Br%C3%A9sil.jpg.

Pg. 55, **Quartz Synthese**: Wikimedia Commons, https://commons.wikimedia.org
/wiki/File:Quartz_synthese.jpg.

Pg. 58, **Police Officer with HT-220**: Motorola, Inc., Legacy Archives Collection. Reproduced with permission.

Pg. 60, **Motorola Pageboy**: Wikimedia Commons, https://commons.wikimedia
.org/wiki/File:Pageboy1.jpg.

Pg. 62, **Motorola HT-220**: Motorola, Inc., Legacy Archives Collection. Reproduced
with permission.

Pg. 65, **Motorola Print Ad**: Motorola, Inc., Legacy Archives Collection. Reproduced
with permission.

Pg. 66, **SCR-536 Diagram**: Motorola, Inc., Legacy Archives Collection. Reproduced
with permission.

Pg. 80, **Bill Weisz, Bob Galvin, John Mitchell**: Motorola, Inc., Legacy Archives
Collection. Reproduced with permission.

Pg. 109, **Flip Mouthpiece Prototype**: Motorola, Inc., Legacy Archives Collection.
Reproduced with permission.

Pg. 109, **Double Flip Prototype**: Motorola, Inc., Legacy Archives Collection.
Reproduced with permission.

Pg. 110, **Retractable Prototype**: Motorola, Inc., Legacy Archives Collection.
Reproduced with permission.

Pg. 110, **Banana Prototype**: Motorola, Inc., Legacy Archives Collection. Reproduced
with permission.

Pg. 111, **Shoe Design**: Motorola, Inc., Legacy Archives Collection. Reproduced with
permission.

Pg. 115, **DynaTAC Sketch**: Don Linder.

Pg. 117, **DynaTAC Patent**: US Patent and Trademark Office 3,906,166.

Pg. 123, **DynaTAC Press Release**: Motorola, Inc., Legacy Archives Collection.
Reproduced with permission.

Pg. 124, **DynaTAC Fact Sheet**: Motorola, Inc., Legacy Archives Collection.
Reproduced with permission.

Pg. 126, **Benjamin Hooks**: Motorola, Inc., Legacy Archives Collection. Reproduced
with permission.

Pg. 144, **Dan Noble**: Motorola, Inc., Legacy Archives Collection. Reproduced with
permission.

Pg. 161, **Mobile and Fixed Phone Adoption in US**: Calculated from data available at Our World in Data, "Technology Adoption." Underlying data from World
Bank, World Development Indicators.

Pg. 167, **Change in Mobile and Fixed Phone Subscriptions**: Calculated from
data in "Leapfrogging: Look Before You Leap," UNCTAD, December 2018.

Pg. 168, **Fixed and Mobile Subscriptions, Low- and Middle-Income Countries**: Calculated from data available at Our World in Data, "Technology Adoption." Underlying data from World Bank, World Development Indicators.

Pg. 176, **NTIA Wall Chart**: National Telecommunications and Information Administration (NTIA), Department of Commerce, 2016.

Pg. 196, **Hearing Aid Collab Diagram**: Harinath Garudadri and Rajesh Gupta,
University of California San Diego.

All other images are courtesy of the author's personal collection. © Martin Cooper.

INDEX